ELECTRICAL ENGINEERING DEVELOPMENTS

LOW POWER HIGH SPEED CMOS MULTIPLEXER DESIGN

ELECTRICAL ENGINEERING DEVELOPMENTS

Additional books in this series can be found on Nova's website under the Series tab.

Additional e-books in this series can be found on Nova's website under the e-book tab.

ELECTRICAL ENGINEERING DEVELOPMENTS

LOW POWER HIGH SPEED CMOS MULTIPLEXER DESIGN

SHYAM AKASHE
AND
KHUSHBOO MISHRA

nova publishers
New York

NOTICE TO THE READER

Library of Congress Cataloging-in-Publication Data

Low power high speed CMOS multiplexer design / editors, Shyam Akashe and Khusbou Mishra.
pages cm. -- (Electrical engineering developments)
Includes index.
ISBN 978-1-63463-322-2 (hardcover)
1. Switching circuits. 2. Mixing circuits. 3. Multiplexing. 4. Electronic circuit design. 5. Metal oxide semiconductors, Complementary. 6. Low voltage integrated circuits. I. Akashe, Shyam. II. Mishra, Khusbou.
TK7868.S9L69 2014
621.3815'37--dc23 2014044402

Published by Nova Science Publishers, Inc. † New York

CONTENTS

Abstract **vii**

List of Figures **xi**

List of Tables **xv**

List of Abbreviations **xvii**

Chapter 1 Scaling Issue of Recent Design of CMOS Device Nanotechnology **1**

1.1. Objective of the Chapter 1

1.2. Innovation of Semiconductor 2

1.3. Early History of Semiconductor 2

1.4. Combinational Circuit 11

1.5. Multiplexer Circuit 12

1.6. Logic Styles 17

1.7. Impact of Logic Style 21

1.8. Proposed Reversible Multiplexer 28

References 30

Chapter 2 Impact of Techniques in Nanosclae CMOS Technology **37**

2.1. Objectives of the Chapter 37

2.2. Technology Scaling and Low Power Circuits 38

2.3. Low Power Digital Cmos Design 39

2.4. Basic GDI Cell (Gate Diffusion Input) Technique 45

2.5. Leakage Reduction Techniques 47

2.6. Different Techniques to Reduce Leakage 49

2.7. Power and Delay Tradeoff in CMOS Circuit 55

2.8. Double Gate Finfet 57

References 59

Chapter 3 Various Logic Style of Multiplexer for Low Power Designs **65**
3.1. Objectives of the Chapter 65
3.2. Multiplexer 66
3.3. Realization of Various Logic Styles of Multiplexer 68
3.4. Reversible Logic 75
3.5. Proposed Reversible Multiplexer 78
References 83

Chapter 4 Leakage Reduction Techniques on Multiplexer Circuit for the System SOC Design **85**
4.1. Objectives of the Chapter 85
4.2. Mechanisms of Leakage Reduction Techniques 86
4.3. Renowned Techniques That Decrease Leakage Power 90
4.4. Body Bias Technique 94
References 97

Chapter 5 Impact of GDI and Finfet on Multiplexer **101**
5.1. Objectives of the Chapter 101
5.2. Introduction 102
5.3. CMOS Based 2:1 MUX 104
5.4. Transmission Gate Based 2:1 MUX 107
5.5. Pass Transistor Based 2:1 MUX 109
5.6. Basic Concept of Gate Diffusion Input (GDI) Cell 110
5.7. Performance Parameter 116
References 119

Authors Contact Information **125**

Index **127**

ABSTRACT

This research proposes the reversible logic Multiplexer and also demarcates between reversible and irreversible logic Multiplexers. For power reduction in future computing technologies, reversible logic is a very productive approach of logic synthesis. In a reversible logic circuit, there is a one-to-one mapping between input and output values where the number of inputs and outputs are equivalent. Reversible circuits are constructed using the reversible gates. The innovative computers erase a bit of information every time they perform a logical operation. In novel design technology this yields to establish the operating frequency and leakage power of reversible logic circuit. This is the inverse correlation between the circuit delay and the leakage current of the reversible circuit. For the width, device smaller than the nominal channel length, the leakage current is increasing exponentially, and time delay of the circuit decreases due to increase of the off-current of the circuit. It describes the various logic gate diagrams of the Multiplexer and the circuit techniques that reduce the leakage variability of the advanced CMOS technologies. This book describes the impact of logic style, that is, how it effects on MUX circuit.

Today's focus on low power is not the only reason for the current growing demands of mobile applications. Even prior to the mobile era, power consumption has been an essential problem. To resolve the power consumption problem, many researchers have proposed divergent thoughts from the device level to the architectural level and above over. Though, there is no universal technique to avoid tradeoffs between power, delay and area, it becomes a requisite for the designers to choose suitable techniques that gratify application and product requirements. The recent enhanced CMOS technology scaling for reversible logic computing capability has increased the feature size

of integrated circuit and miniaturized the transistor in the reversible logic gate low power with garbage input/output with a design circuit that is 2:1 reversible multiplexer. Presented here is the functionality of the circuits' frequency response of the logic gate operation of 2:1 reversible multiplexer which is increasing the integrated circuits with each new generated technology by enhancing the future technology. This book also describes the various leakage reduction techniques in CMOS Multiplexer design.

For low-power, Very Large Scale Intigration (VLSI) can be assigned to special design levels, such as the behavioural, structural, layout and the process technology level are the escalating requirements. Multiplexer is the primary component which consists of CMOS (Complementary Metal Oxide Semiconductor) memory elements and data control structures. The low-power consumption is one of the most significant concerns of the system design. In high speed applications, different styles are developed for low power designs.

The purpose of this book is to reduce power and area of 2:1 MUX, 4:1 MUX and reversible logic while maintaining the viable performance. The diverse configurations are designed using different topologies of 2:1 MUX and 4:1 MUX such as CMOS based MUX, transmission gate and pass transistor. Multiple logic styles are used to design a low power 2:1 MUX and 4:1 MUX. The leakage current, power consumption, delay and transistor count up is compared for different logic styles of 2:1 MUX and 4:1 MUX. Reversible logic has emerged as one of the most important approaches and more prominent technology having its applications in low power CMOS, garbage inputs/outputs, communication, nanotechnology and computer graphics. This chapter presents a novel reversible multiplexer gate and the design of differential reversible multiplexer using the proposed reversible gate is discussed.

Technology scales and leakage currents grow exponentially and become an increasingly large component of total power dissipation. In this book, influence of leakage reduction techniques on multiplexer/load circuit is analyzed. The low-power consumption is one of the most important issues for the system SOC design; therefore, different techniques and technologies for low-power designs in high speed interface applications are developed and also applied in the practical design projects. In order to suppress the degradation of signals and to increase the operation speed, we designed interconnection for the circuit using Self-controllable Voltage Level (SVL) techniques. The designed circuit is realized in a high variability standard 45nm process and uses 0.7V supply voltage. Our optimization technique, using the proposed method, reduces power consumption and leakage current by significant

amount of multiplexer /load circuit. Body bias technique which decreases the process variation on MUX is the key applied on MUX to decrease low power, leakage current and bulk voltage. We investigate the use of body biasing to adaptation control the bulk voltages of transistor and exhibit that it is particularly effective in sub- threshold circuits and can eliminate performance variations with low power.

The intention of this book is to reduce power and area of 2:1 MUX while maintaining the viable performance. The diverse configurations are designed using different topologies of 2:1 MUX such as the topology of MUX was designed using CMOS, transmission gate and pass transistor using FINFET(Fin shaped Field Effect Transistor). The mobility was improved in devices with taller fins due to increased tensile stress. To develop performance and reduce power consumption, DG FINFET can be proficiently used. In non-critical paths, sovereign gate control can be used to join mutually parallel transistors.

In this book, the design of proposed reversible logic multiplexer with garbage input and output, Gate -Diffusion -Input (GDI cell) technique are used in circuit realization, in which there is low power and area design in this dynamic constituent of condensed power. In GDI cell, PMOS is not linked to supply voltage VDD and NMOS is not linked to the GND. GDI circuits present some major issues of enhanced hazard tolerance and low voltage operation of reversible logic circuits. In this book, it is proposed a new application of GDI (Gate-Diffusion Input) circuits to Reversible logic multiplexer with its Garbage input and output. The novel proposed design technique will consume less power than the other conventional gate. Reversible logic circuit has displayed less power dissipation in recent years. This GDI cell technique decreases the power of the circuit, delay, Power-Delay Product (PDP) and it also compacts the frequency. The device scaling is partial as the power dissipations is more optimized in terms of delay, frequency jitter, bandwidth power supply, frequency and duty cycle of the signal and also establishes the noise of the circuit. In the reversible logic design, the GDI is efficient in lower delay, low power and low leakage current.

LIST OF FIGURES

Figure 1.1: The minimum feature size and year wise down scaling technology 6
Figure 1.2: Total chip power based on ITRS roadmap in 2004, we are just breaking even [Kim, et al., Computer 2003] Scalable X86 CPU Design for 90nm Low V_T devices are <1% of total non-memory transistor width (Leakages and Dynamic Power Dissipation) [J. Schultz and C. Webb, ISSCC 2004] 7
Figure 1.3: Combinational Logic 11
Figure 1.4: Classification Combinational Logic 14
Figure 1.5: Block Diagram of MUX 14
Figure 1.6: Block Diagram of 2:1 MUX 14
Figure 1.7: Block Diagram of 4:1 MUX 15
Figure 1.8: Symbol Diagram of 2:1 MUX 16
Figure 1.9: Gate level Diagram of 2:1 MUX 16
Figure 1.10: Block diagram of the 2:1 MUX IC. 17
Figure 1.11: Simple block diagram of 2:1multiplexer 18
Figure 1.12: Gate level design of multiplexer 18
Figure 1.13: Gate level Mux design for the all level of gate 19
Figure 1.14: Transmission gate 19
Figure 1.15: Transmission gate Multiplexer 20
Figure 1.16: Simple 2:1 multiplexer 21
Figure 1.17: A gate with K input and Output is call K*K gate 25
Figure 1.18: Combinational circuit diagram of 1x1 NOT Gate and CNOT Gate 27
Figure 1.19: Basic block of reversible logic gate with Garbage input/output 27

Figure 1.20: Implementation of reversible multiplexer with garbage input/ output Gate (a) Circuit diagram for multiplexed output Y (b) Circuit diagram for output G2 (c) Circuit diagram for output G1 29
Figure 2.1: Impact of Technology on total chip power 38
Figure 2.2: Dynamic Power Consumption 41
Figure 2.3: (a) short-circuit power dissipation; (b) short-circuit power with large capacitive load, (c) short circuit power with small capacitive load 43
Figure 2.4: Leakage (static) Power consumption 44
Figure 2.5: Basic Gate-Diffusion Input cell 46
Figure 2.6: Leakage current component in a transistor 50
Figure 2.7: MTCMOS technique structure 51
Figure 2.8: Transistor stacking effect 51
Figure 2.9: MOSFET as a three-terminal device (a) n-MOSFET (b) p-MOSFET. Here, S, G, D and B, respectively indicate source, gate, drain and substrate. 53
Figure 2.10: CMOS inverter under anti body-source short 53
Figure2.11: Two level multiple static supply voltage schemes. 54
Figure 2.12: Schematics of the fabricated FINFETs taking Hfin as a parameter 58
Figure 2.13: Simulated FINFET cross section 58
Figure 3.1: Graphical Symbol 4:1 MUX. 66
Figure 3.2: Gate implementation of a 4:1 Multiplexer. 67
Figure 3.3: Graphical Symbol 2:1 MUX 67
Figure 3.4: Static CMOS 4:1 MUX 68
Figure 3.5: Schematic of CMOS based 2:1 MUX 69
Figure 3.6: Transmission gate based 4:1 MUX 71
Figure 3.7: Schematic of Transmission gate 2:1 MUX 71
Figure 3.8: Pass-Transistor 4:1 MUX 72
Figure 3.9: Schematic of pass transistor 2:1 MUX 73
Figure 3.10: Pseudo NMOS Transistor 4:1 MUX 74
Figure 3.11: LEAP 4:1 MUX 75
Figure 3.12: Reversible Inverter 78
Figure 3.13: CMOS Transmission gate 78
Figure 3.14: 2:1 Multiplexer uses Pass Transistor 79

Figure 3.15: Implementation of reversible multiplexer with garbage input/ output Gate (a) Circuit diagram for multiplexed output Y (b) Circuit diagram for output G2 (c) Circuit diagram for output G1. 80
Figure 3.16: Transistor implementation of 2:1 reversible multiplexer with Garbage input/output 81
Figure 3.17: Delay comparison 2:1 reversible multiplexer with garbage input/output 82
Figure 4.1: MTCMOS 87
Figure 4.2: VTCMOS 88
Figure 4.3: Path distribution of Dual Vth 89
Figure 4.4: Upper SVL Circuit 91
Figure 4.5: Lower SVL Circuit 92
Figure 4.6: SVL Circuit 93
Figure 4.7: MOSFET as a three-terminal device (a) n-MOSFET (b) p-MOSFET. Here, S, G, D and B, respectively indicate source, gate, drain and substrate. 96
Figure 4.8: CMOS inverter under (a) body-source short and (b) forward-body bias conditions. The VBS, N is the forward body-bias across the n-MOSFET and VSB, P is the forward body-bias across the p-MOSFET 97
Figure5.1: Symbol Diagram of 2:1 MUX. 105
Figure 5.2: Schematic of CMOS based 2:1 MUX 105
Figure 5.3: Schematic of CMOS based 2:1 MUX using FINFET 106
Figure 5.4: Schematic of Transmission gate 2:1 MUX 108
Figure 5.5: Schematic of Transmission gate 2:1 MUX using FINFET 108
Figure 5.6: Schematic of pass transistor 2:1 MUX 109
Figure 5.7: Schematic of pass transistor 2:1 MUX using FINFET 110
Figure 5.8: Basic Gate-Diffusion Input cell 111
Figure 5.9: Combinational circuit diagram of 1x1 NOT Gate and CNOT Gate 113
Figure 5.10: Transistor implementation of 2:1 reversible multiplexer with Garbage input/output using GDI Cell technique 115

List of Tables

Table 1.1: Oxide thickness scaling prediction Per the ITRS 2006 update, manufacturing solutions for metal-gate insulators at 32 nm and beyond are not yet known, and at the 16 nm node are presently undefined. 9
Table 1.2: Truth Table for Reversible EXOR gate logic 26
Table 2.1: Functional table for Basic GDI cell 47
Table 3.1: Functional Table of 2:1 Reversible Multiplexer 79
Table 3.2: Truth Table of 2:1 Reversible Multiplexer with Garbage Input/output 81
Table 5.1: Truth Table of the basic GDI cell (Gate Diffusion Input) 112
Table 5.2: Truth Table of 2:1 Reversible Multiplexer with Garbage input/output using GDI technique 115

LIST OF ABBREVIATIONS

Abbreviation	*Meaning*
VLSI	Very Large Scale Integration
CMOS	Comlentary Metal Oxide Semiconductor
GDI	Gate Diffusion Input
MUX	Multiplexer
FINFET	Fin Shaped Field Effect Transistor
MOS	Metal Oxide Semiconductor
FET	Field Effect Transistor
BJT	Bipolar Junction Transistor
CMOS	Complementary Metal Oxide Semiconductor
nMOS	n – Channel MOSFET
pMOS	p – Channel MOSFET
ns	Nano Second
ps	Pico Second
nW	Nano Watt
mW	Milli Watt
μW	Micro Watt
MTCMOS	Multi-Threshold Complementary Metal Oxide
VTCMOS	Variable Threshold CMOS
VDD	Drain to Drain Voltage
GND	Ground
DCVS	Differential Cascade Voltage Switch
SCE	Short Channel Effect
DIBL	Drain Induced Barrier Lowering
GIDL	Gate Induced Drain Lowering

Vds	Drain to Source Voltage
Vgs	Gate to Source Voltage
Ids	Drain to Source Current
TGLGATE	Transmission Gate Logic Mux
PTL	Pass-Transistor Mux

Chapter 1

SCALING ISSUE OF RECENT DESIGN OF CMOS DEVICE NANOTECHNOLOGY

1.1. OBJECTIVE OF THE CHAPTER

This chapter proposes the reversible logic Multiplexer and it also differentiates between reversible and irreversible logic multiplexer. For power reduction in future computing technologies Reversible logic is a very prospective approach of logic synthesis. In a reversible logic circuit, there is a one-to-one mapping between input and output values the number of inputs and outputs are equivalent. Reversible circuits are constructed using reversible gates. The innovative computers erase a bit of information every time they perform a logical operation. In novel design technology this is yields to establish to the operating frequency and leakage power of reversible logic circuit. There is an inverse correlation between circuit delay and leakage current of reversible circuit. For the width device smaller channel length than the nominal, the leakage current is increasing exponentially, and the time delay of the circuit is decreases due to increase of the off current of circuit. It describes the various logic gate diagram of the Multiplexer. The circuit techniques that reduce the leakage variability of advanced CMOS technologies have been described through the various logic gate diagrams of the Multiplexer. This chapter describes the impact of logic style, i.e. how it affects the MUX circuit.

1.2. Innovation of Semiconductor

1.3. Early History of Semiconductor

Most of the history of semiconductor is long and complicated. Obviously, it is difficult to fit it in one short thesis organization. The given limitation to the Authors concentrated on the fact that is considered the most important, so this choice is never fully impartial. Therefore, various apologies to this important research work to all those readers who will find that some important and vital moments of history of semiconductor industry are missing.

According to G. Busch the word "Semiconducting" was used first time by Alessandro Volta in 1782 [1]. The first observation of semiconductor effect is introduced by Michael Faraday (1833), who noticed that the resistance of silver sulphide decreased with the temperature, which was different than the other dependence in metal material [2]. An extensive quantities analysis of the dependence on temperature on electrical conductivity of Ag2S and Cu2S was published in 1851 by Johann Hittorf.

In the 19th Century i.e. in the year 1974, came the most significant discovery in the field of semiconductor. It leds to the discovery of rectification during the contact between various materials, that is especially, naturally, occurring sulphide crystal material. Braun's discovery was mostly related to natural crystal and Schuster's discovery to another contact between tarnished and untarnished copper wire that is used for the semiconducting material [3]. In Schuster's experiment, the layer of copper oxide is untarnished wire which presumably acted as the semiconductor giving the contact of a rectification property [4].

In the year of 1878 Edwin Herbert Hall discovered that the charge carrier of the solids deflected the magnetic field (Hall Effect) that was produced in the circuits. This basic phenomenon is used to study various properties of semiconductor [5]. After the discovery of electron by J.J Thomson, several scientists proposed the theories of electron based metal conduction of material.

1.3.1. Effect on Computing

The history of electronics industry is really a vast area and it is not possible to present the whole detail of the semiconductor history as is limited to this thesis organization. They started of with electronics philosophy, then

physics and then electrical engineering which has now got its own identity and going to be more diverse as the future scope. There is no doubt about it that the modern electronics as we see it today started from the birth of vacuum diode of Sir Ambrose Fleming. During the centenary year, we should remember the great personalities as his predecessors and successors. The 20th century is mainly running due to electronics.

The first general purpose electronic computer was introduced by John Mauchly and John Presper Eckert of the University of Pennsylvania in 1946[6], Electronic Numerical Integrator and Computer (ENIAC). The first Electronic computer, ENIAC, was jubilated as a "Giant Brain" in the press [7]. That aroused a keen interest among the scientists and industrialists [8]. Electronic Numerical Integrator and Computer (ENIAC), with 17,468 vacuum tubes, was the pinnacle in the era of vacuum tubes. Thereafter, the volume and power consumption of ENIAC astonished the CMOS VLSI design industry [7]. The weight of ENIAC was more than 27 tons and it occupied $167m^2$ area [8]. The highest most power consumption of the ENIAC was 150kW [7], the same level of power consumption of today's advance technology data canters [8].

In the year 1947, the three Physicists William Shockley, John Barden and Walter Brattain of Bell Labs fabricated the first working transistor to replace the bulky and the unreliable vacuum tubes [6]. The first transistor consisted of germanium base with two closely gaped gold contacts which are held by spring. Shockley, Bardeen, and Brattain observed that the small applied voltage on one of the contacts modulates the current flow between the other two terminals of the vacuum transistor [9]. That's amplifying the input signal by more than 100 times [6]. The transistor is one of the most considerable inventions of the 20th century for the semiconductor industry [9].

The most predicted silicon-based transistors require a high-purity of silicon substrate. Alternatively, the requirement of the purity of germanium-based transistors is relatively lower [6]. The Germanium transistor is, therefore easier to process as compared to the silicon material [10]. Furthermore, germanium materials provide a better conductivity with the higher electron and hole motilities' as compared to the silicon material [6]. The first most transistor was fabricated with germanium material [9]. The application of the germanium material devices is limited due to the higher cost of germanium as compared to silicon material devices [10]. Morris Tanenbaum of Bell labs managed the fabrication of first silicon material based transistor in January 1954 [6]. The Bell labs considered that the silicon transistor is not attractive for commercial application and did not focus on the research. In May 1954,

[10] Texas Instruments Incorporated (TI) announced the invention of the first silicon material transistor and pushed the silicon transistor into a low cost level commercial production of competition of semiconductor industry material [10].

The solid state transistor occupied drastically smaller volume and consumed lower power as compared to vacuum tubes transistor. So more/less the numbers of transistors are required to design more advanced technology based electronic systems with enhanced functionality and performance of the transistor. It became a great challenge to pose the discrete transistor interconnections. In September 1958, Jack Kilby of Texas Instruments Incorporated built the first integrated circuit (IC). Kilby integrated transistor resistor and capacitors on a single germanium base. This used "flying wires" which were made of the gold. Kilby integrated all the components interconnected together. Since the wires were off-chip, still Kilby Integrated circuit faced the reliability issues. Robert Noyce of Fairchild introduced semiconductor technology late in the year 1959.

Semiconductor gave the idea of monolithic integrated circuit [6]. In the late 1960s, Dawon Kahang and Martin Atalla of Bell labs exhibited the first Metal-/Oxide-Semiconductor Field Effect Transistor (MOSFET) [9]. This technology based new transistor is called as the MOSFET. C.T. Sah and Frank Wanlass of Fairchild semiconductor proposed the Complementary Metal Oxide Semiconductor (CMOS) logic Family circuit in 1963 [6]. The most CMOS circuits use less area, consumes less power and display significantly higher noise Immunity of the logic circuit as compared to the bipolar transistor, PMOS and NMOS circuits [10]. The CMOS scaling technology become preferred for design and manufacturing of semiconductor in 1980s.

In 1971, the Intel Corporation designed the first microprocessor, 4004[6]. The microprocessor 4004 contained 2300 transistors that were fabricated in 10 μm PMOS technology. The required clock frequency of 4004 was 108 kHz. This provided similar computational performance with smaller area, lower power consumption and enhanced significance of reliability issues as compared to ENIAC [10]. In the year 2011, after four decades of technology scaling, one of the most power full microprocessors was released by the Intel Corporation, Itanium Poulson, containing 3.1 billion transistors fabricated in 32nm bulk CMOS VLSI technology [11]. In the year 2011, Intel Corporation developed a new silicon transistor structure (tri-gate bulk MOSFET) for transition to the 22nm CMOS technology node [12].

One of the most significant issues in today's latest advance low power ideas of last century; the Complementary Metal Oxide Semiconductor

(CMOS) technology [13]. It has served us well for various decades. However, to make the best performance of the cost in trade-offs design of circuits [14], we continue this basic concept of integrating onto a chip by shrinking and using the feature of the small size of chip that is permitted by the manufacturing technology. As a result, the dissipation of the power in CMOS VLSI per unit area grows and that solves the heat removal problems and cooling issues. Varied and several examples in this context are the general-purpose microprocessor used in desktop computers, laptops and mobile phones. Even when scaling technology down to the supply voltage, power dissipation does not come down. Today, the more important issue of reversible logic circuit is the scaling property. The down scaling size of the semiconductor material is also required in less amount in design of VLSI circuits.

1.3.2. Nanotechnology: A Brief Overview

The trend of Complementary Metal-Oxide-Semiconductor (CMOS) is Nanoscale technology, evolutionary exponential, predicted by Moore's law and has been followed seemingly under the demand of semiconductor industry. The growth of semiconductor industry is based on better performance and it faces fierce global competition over the ages. The driving force behind this is that the very swift progress for long-term commitment to steady downscaling Nano technologies is needed to meet the specific requirement on speed, area, complexity, circuit density and the power consumption posed by many of advanced applications of the last year. The degree of scaling is measured as the half-pitch of the first level of interconnects in the DRAM technology. The degree of Nanoscale is also termed as the "Technology node" by the International Technology Roadmap for Semiconductors (ITRS). According to the update of ITRS 2006, the 2007 Production Technology node has reached through these years as 45 nm, while the smallest features size of the transistor, the effective MOS field-effect transistor alters the various gate properties like the gate length, area, power and low leakage as merely 22 nm technology. These numbers are expected to approach 11 and 7 nm, respectively during the next ten years [15].

In The Introduction, I motivated research in Nanoscale or Nanotechnology by taking a look at the nature. The forecaster of industry is care about next quarter's returns and less about the theoretical limits of nanotechnology.

I provide here the basic motivational concept for Nanotechnology: Moore's Law.

1.3.3. Moore's Law

In 1965, Gordon Moore made the prediction, that the number of transistor count is doubled every 18 to 24 months, implying an exponential rise in computer power a year ago, but soon it was found that it was amazing and accurate as today's technology scaling. It is essential from microelectronics industry point of view. It is also assumed that the power is doubled every year and half the investors are able to justify this.

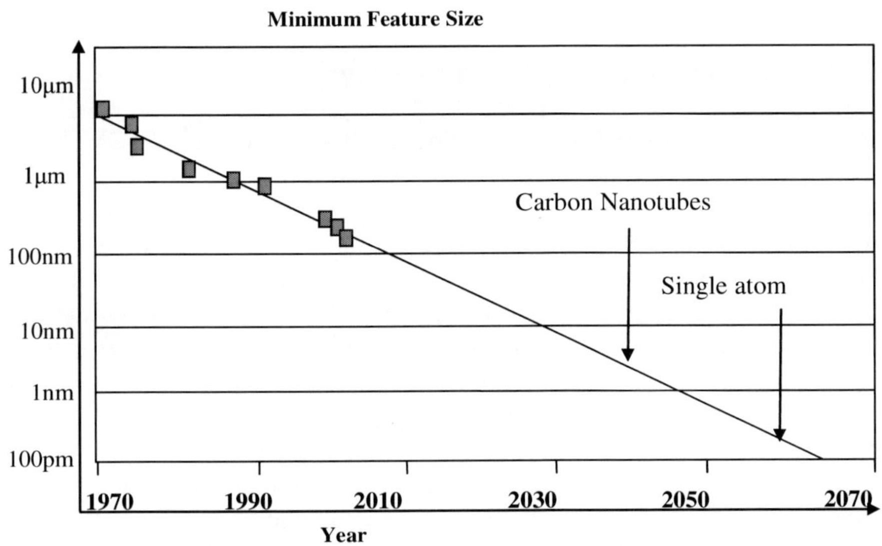

Figure 1.1. Minimum Feature Size and Year Wise down Scaling Technology.

So far, the Semiconductor industry has kept up with the Moore's law prediction as shown through black line. The problem, however, is kept that in the future the demand will be much greater. By 2010, we will need to define the size of Nanoscale technology is 50nm. By 2040, wire line widths will be reaching on the order of atoms of carbon nanotubes, and by over year 2060, features will be the size of individual transistor like atoms. So, it seems that it will be physically impossible to continue hold on Moore's law beyond the

single-atom barrier, and the other fact is that the physical constraints are likely to block the progress much sooner. The modern fabs cost billions to build and run, mostly due to the stringent clean room condition which requires whole chip fabs in features size in 100 nm across. As in figure 1.1, the design of the figure displays the minimum feature size of the down scaling technology.

The prime purpose of presenting the Moore's law is not to scare the reader into believing that the semiconductor industry and the computer industry will collapse in the near future. It is truly said that the computer is not more powerful than the preceding prediction. It is just that the industry will have to adapt a new technology day by day for next generation computers. In other words it simply says that in short terms by adopting these new micro-and Nano fabrication techniques, it is essential to build the conventional CMOS transistor is too small. In long terms, this means adopting another way of new fabrication techniques. In this case, more research in Nanotechnology is key for feature size for the transistor.

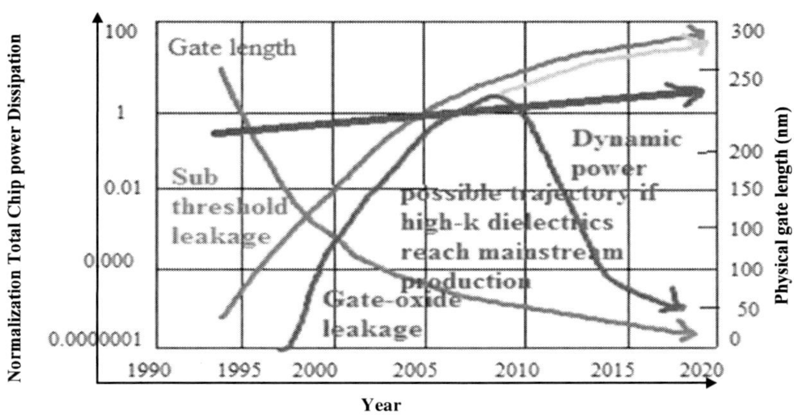

Figure 1.2. Total chip power based on ITRS roadmap in 2004, we are just breaking even [Kim, et al., Computer 2003] Scalable X86 CPU Design for 90nm Low V_T devices are <1% of total non-memory transistor width (Leakages and Dynamic Power Dissipation) [J. Schultz and C. Webb, ISSCC 2004].

Figure 1.2 shows that the data point for the number of transistors is increasing over the years. Clearly Moore's law held that for many years. But the question arises. "How the semiconductor industry is to hold that for a long decade?" The first and the foremost problem are to increase the number of transistor count in integrated circuits per chip, so that the size of the transistor

is smaller and smaller. This means that the fabrication or manufacturing process must be able to diminish the feature size over the years.

With the introduction of Nanoscale CMOS (Complementary Metal-Oxide Semiconductor) technology (sub-100 nm), today digital and analog design confronts many new challenges at different level of places in integrated circuits. These challenges include several degradations according to device matching characteristics. As a result of lithographic quantum limits, the resolution of enhanced technique results in the device image on the reticle's, no longer having a one-to-one correspondence with the design of layout because the feature size of transistor is less than one-quarter of the wavelength of deep ultraviolet rays (UV). Illustrating this, the high leakage currents result from the low level of threshold voltage and thin gate oxide tunneling currents, thus electro migration limits require more parallel level of the metal interconnect shared-drain analog devices, and the limitation on accurate pre-layout simulation of circuit design. The device counts are the greater than 100 million on a single chip, driving the new methodology for designing the partitioning system power densities.

1.3.4. Nanoscale CMOS Physical Design and Device Trends

According to the International Roadmap of Semiconductor (ITRS) the power and variability are the two top most challenges which act as obstacles in the design for Integrated Circuits (IC) [16]. The leakage power is increasing exponentially in each technology node. This region is due to the 15% reduction of the threshold voltage in each scaling technology node in semiconductor which increases the sub-threshold leakage by about five transistors. This is expected at the 45 nm technology node, with the leakage power of 50% of the total consumption power of portable devices application. The aggressive scaling technology of oxide thickness to control the short-channel effects, leads to the significant tunneling gate leakage component i.e. to include the device off-current [17].

The ITRS working group has published many designs and device targets for Nanoscale technology of nodes for the next decade. There are many device sizes and designs but the rule remains that relatively constant fraction of the gate length is minimum in transistor sizing. This is passed to several past and future process Nanoscale technologies. This is the downscaling issue of recent design technology. ITRS technology is for the reversible logic based

semiconductor industry in the design of recent technology in design of logic circuit technology.

Table 1.1. Oxide thickness scaling prediction Per the ITRS 2006 update, manufacturing solutions for metal-gate insulators at 32 nm and beyond are not yet known, and at the 16 nm node are presently undefined

Node F (nm)	Year	ITRS T_{OX} HP	ITRS T_{OX} LP	0.02*F T_{OX} (nm)
65	2007	1.30		1.30
45	2010	0.90	1.40	0.90
32	2013	0.50	1.20	0.64
22	2016	0.50	0.60	0.44
16	2019	?	?	0.32

LP devices do not follow HP scaling trends. LP devices Provide higher values of Tox, and therefore higher operating voltages.

In table 1.1, ITRS SiO2-equivalent gate oxide thickness (Tox) targets for high-performance (HP) of device as compared to the column 3 of the table with low power (LP) devices in column 4 and general prediction of the oxide thickness. At the Nanoscale technology node beyond t 45 and 22 nm level, the gate leakages, as a result of direct tunneling, plays a significant role in device downscaling technology. At the 100 nm technology and beyond, the thickness of gate oxide is realized value of high dielectric constant.

The increased variability in recently used advanced CMOS technology plays a vital role to determine the total leakage of chip. This has accomplices to need the account for statically leakage variation during the design cycle of integrated circuit. It is very difficult to achieve the required speed while meeting the power constraints, because of excessively large spread in currents.

The whole design of chip, is because of large variation in leakage power as a high twenty transistor for 30% variation in its frequency. The required highest frequency chip has wide distribution of leakage, and for the given leakage there is so much of distribution in frequency of the chip [18].

In new design technology, this yields to determine the operating frequency and leakage power of reversible logic circuit. It has been shown that among the chip to meet the required operating frequency and large fraction of much amount of leakage power, makes them unsuitable for usage and degrades the result [19]. This is the inverse correlation between circuit delay and leakage current of reversible circuit. For the device with smaller channel length than

the nominal, the leakage current is increasing exponentially, and time delay of the circuit decreases due to increase in the off-current of the circuit. Therefore, higher the chip frequency response, the more it suffers from large leakage power in the circuit. There is crucial need for accurate estimation and prediction of leakage current which is account statistical variation. The circuit techniques reduce the leakage variability of advanced CMOS technologies.

1.3.5. Nanoscale CMOS Reliability Issues

Many aspects of device circuit parameter and supply 1 V voltage scaling was predicted accurately by Dennard et al. in 1974[20]. Changes in device and performance of the circuit were indicated in terms of dimensionless scaling factor k. The prediction of constant power density was apparently based on the static current of the reversible circuit. The supply voltage and static current both scale level as 1/k. The scaling technology also depends on the threshold voltage of the device. This CMOS device scaling issue is recent design of Nanotechnology. These times are one of the best ones to use the scaling device technology for higher design implementation of reversible logic gate.

The recent trends in digital CMOS having independents IP provider's favours more in circuit design activities fables centers that have the limit of reliability of Nanoscale technology in engineering capability. Unfortunately, many of the founders of semiconductor industries reported that the reliability stress limits are documented in terms of digital, not to analog, circuit end of life. The gap in recent result that is predicted here required gap between available device reliability data and simulation tools. Despite the efforts of the industry to narrow as is called "the Reliability gap", it is widening with time [21]. There are number of design techniques to design integrated circuit employed to address the major non-idealities of progress of Nanoscale technology, when this approach is using to design SOC (System-on-chip) chip. The most non-idealities, in order of the poor predictability of progress parameter, determine the impedances and capacitance. The constant current approach is based in system on chip establishing accurate and temperature sensitive in bias currents [22]. This approach is based on applying an accurate band gap voltage across an accurate off-chip resistor design; this is due to generate the accurate current that is copied and sent to the various on-chip blocks.

With the parallel current sources and the current mirror, the second most alternative approach is similar but uses a system-on-chip resistor rather than an on-off resistor.

1.4. COMBINATIONAL CIRCUIT

These circuits are contrasting Sequential Logic Circuits whose outputs are reliant on both their present inputs and their preceding output condition giving them various forms of memory. The outputs of Combinational Logic Circuits are only resolute by the logical function of their current input state, logic '0' or logic '1', at any particular moment in time as they have no feedback, and any changes to the signals being useful to their inputs will instantaneously have an effect on the output.

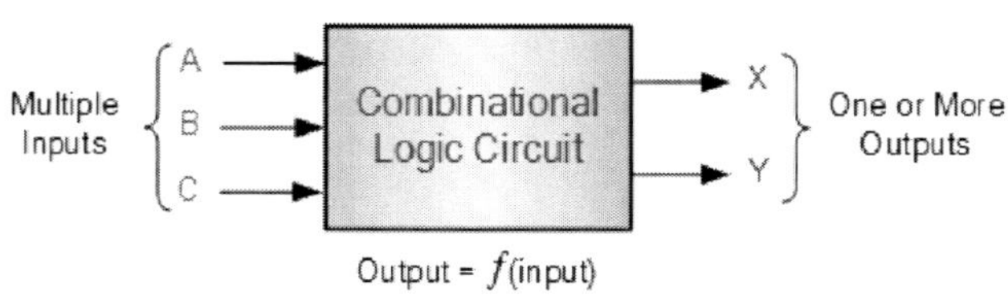

Figure 1.3. Combinational Logic.

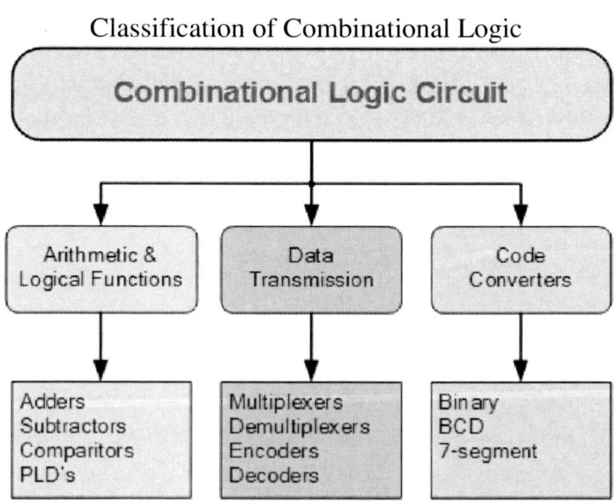

Figure 1.4. Classification of Combinational Logic.

Further, in a Combinational Logic Circuit, the output is dependent, relatively at all times, on the combination of its inputs and if one of its inputs changes state so does the output as combinational circuits have 'no memory', 'timing' or 'feedback loops'.

Combinational Logic Circuits are prepared from basic logic NAND, NOR or NOT gates that are 'mutual' or coupled together to produce more complex switching circuits. Combinational logic circuits are the building blocks of these logic gates. Then can be very simple or very complicated and any combinational circuit can be implemented with only NAND and NOR gates as these are classified as 'universal' gates. The basic three main behaviors of specifying the purpose of a combinational logic circuit are:

- Boolean algebra-This forms the operation of the logic circuit showing the algebraic expression outcome for each input variable as True or False in a logic '1' output.
- Truth Table-It defines the function of a logic gate by providing a concise list that shows all the output states in tabular form for every probable combination of input variable that the gate could meet.
- Logic Diagram-It is a graphical representation of a logic circuit that presents the wiring and connections of each entity logic gate and the logic circuit that implement the specific graphical symbol.

Individual use of combinational logic is in multiplexer and de-multiplexer type circuits. Here, multiple inputs or outputs are attached to ordinary signal line and to decode an address, to select a single data input or output switch, logic gates are used. A multiplexer contains two split components, a logic decoder and several solid state switches, but prior to conferring the multiplexers, we first need to understand how these devices use 'solid state switches' in their design decoders and de-multiplexers in more detail.

1.5. Multiplexer Circuit

In any arithmetic circuit, multiplexer is the heart of the circuit abbreviated as MUX. It is the vital part of computing devices starting from convenient devices to supercomputers and more. Mainly, its role in diminutive devices is an important one. The battery need for long battery back-up time, for the diminutive devices, is a key concern which always enhances the expectations

of the users for more and more backup time. Hence, the power consumed by multiplexers is a key issue to manage. In a data path, the multiplexer with more bits is replaced by 2-1 MUX as a hierarchy called decomposition. There are diverse approaches followed for power optimization in a decomposed MUX hierarchy [23]. They frequently aim to ease power consumption and enhance speed. These papers have also investigated diverse approaches implementing MUX using CMOS pass transistor; each has its pros and cons. There is the tradeoff present between supply voltage, current, power, area and delay. Therefore, only some recital criteria will be assembled at the same time. Diverse approaches have been proposed to decrease power consumption of MUX hierarchy. Several researches covenant it at the algorithm level and various at the circuit level [24]. An n-to-one MUX is distorted into an equivalent hierarchy of two-to-one MUX. There are diverse combinations which are possible to build. But it is required to use the proper MUX structure to do so for power optimization. Even when in non portable applications power is available, the concern of low-power design is fetching decisive, which ensures all the factors to be sacrificed rather than fulfilled because of the trade- off discussed a little while ago. This causes everyone to think about the careful design. In this way, we need low power as well as speed for our design. In this research, the proposed style is compared with Static CMOS and Pass transistor (transmission gate) type of MUX hierarchy. In a number of applications, data paths and data-switching structures are used extensively including processors [25], processor buses, network switches and DSPs with resource allocation. By far, outweighed, the area and power of functional units and registers have reconfigurable architectures such as FPGAs, the area and power of MUX and interconnect. For example, it has been predicted that MUX typically accounts for over 25% of the area of an FPGA design, in the Altera Benchmark set of 120 real customer designs. Further, in [26], CMOS technology in FPGA designs, it was shown that the delay and power data of a 32-to-1 MUX are almost equivalent to those of an 18-bit multiplier in 0.1μm [27]. Consequently, the optimization of MUX is at the heart of digital architecture recital tuning in conditions of power, area and speed. Diverse approaches have been proposed to decrease power consumption and area in MUX hierarchy, both at the algorithm level [28] (particularly as concerns testing, diagnostics and optimization of a vibrant reconfigurable FPGA) and at the circuit level [29] (by significant the suitable logic styles to equivalent the application type). The use of promising technologies, and exclusively 4-terminal devices [30], can also be a capable alternative track for this concern since such approaches have earlier resulted in highly proficient solutions in

period of power consumption [31], area and delay when used for customary or reconfigurable cells [32].

In VLSI design, multiplexer is a basic element that is frequently used. In conditional statements of codes written in hardware description language to control the path of data movement from two or more operands to a shared operator or to synthesize, multiplexer is broadly utilized in diverse design blocks. In data path logic to provide multiple connections between functional units (FUs) and registers, a multiplexer (MUX) is a standard complex gate often used. With its input number, the multiplexer area is not linear. To drive the MUX output, a k to 1 MUX selects one of the k inputs of the MUX.

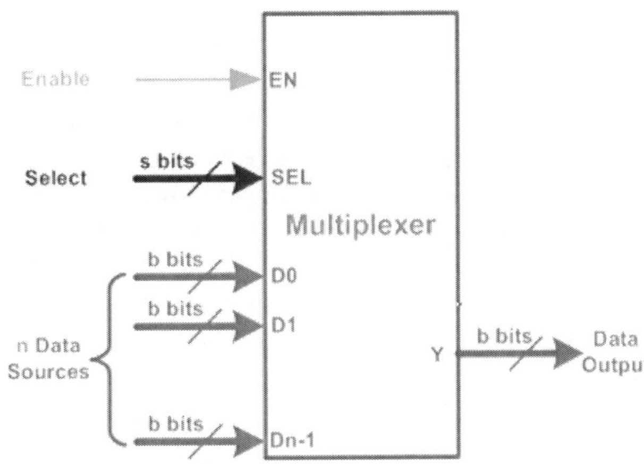

Figure1.5. Block Diagram of MUX.

A multiplexer through two data inputs is referred as '2-to-1 or 2:1' MUX. Frequently used circuit and graphical symbol for 2:1 multiplexer is represented in Fig1.6. Correspondingly, multiplexer with four data inputs is referred to as 4:1 multiplexer. Block diagram of 2:1 MUX is shown in figure 1.7.

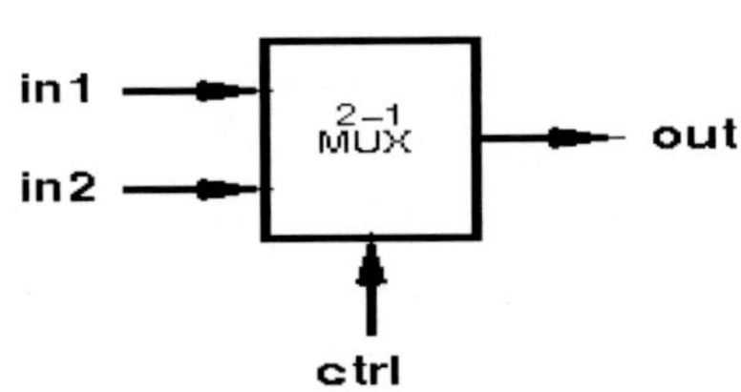

Figure 1.6. Block Diagram of 2.1 MUX.

A sum-of-products (SOP) implementation of the 4-to-1 multiplexer expression is displayed below. It realizes the multiplexer function

$$\text{Output}=X0.\overline{C_0}.\overline{C_1}+X1.\overline{C_0}.C1+X2.C0.\overline{C_1}+X3.C0.C1 \qquad (1.1)$$

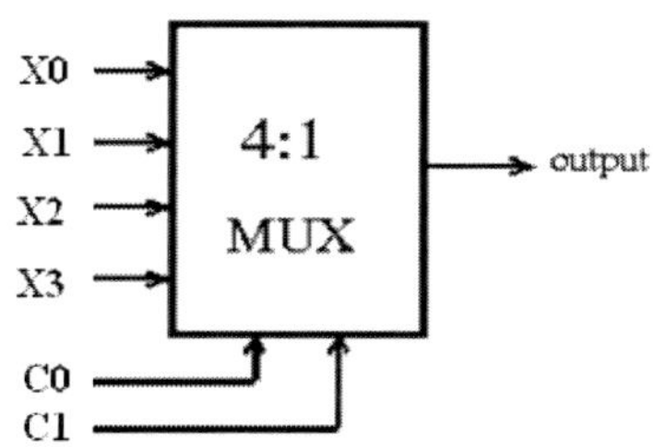

Figure 1.7. Block Diagram of 4:1 MUX.

Suppose we have four lines, X0, X1, X2 and X3, which are to be multiplexed on a single line, Output (M). The four input lines are also identified as the Data Inputs. Since, here are four inputs, we will need two extra inputs to multiplexer, identified as the Select Inputs, for which of the X inputs is to appear at the output to select, and called as select lines C0 and C1. The gate realization of a 4:1 MUX is shown in figure 1.8. Equation 1 is given for 4:1 MUX.

1.5.1. Multiplexer 2:1

Frequently, n-to-1 multiplexers are realized through decomposition into 2-to-1 multiplexer. Because the hierarchy and the selection signal order of a multiplexer hierarchy can affect the quality of a design, the correlated researches on n-to-1 multiplexer decomposition problem have been dedicated to minimize design considerations such as timing, area and power consumption. So, distant multiplexer decomposition procedure for low power have mainly been based on the evaluation of the number of signal transitions modeled with on-probability of data and selection signals. Though, estimation of power using on-probability is not exact because it does not replicate accurate amount of signal transitions.

A straightforward implementation of 2-to-1 MUX would require 2 AND gate, an OR gate and a NOT gate. Symbol diagram of 2:1 MUX is shown in

fig. 1.9 and Gate level 2:1 MUX is shown in fig. 1.10. Any logic function in complementary CMOS is implemented by NMOS pull-down and PMOS pull-up networks attached between gate, output and power lines [33].

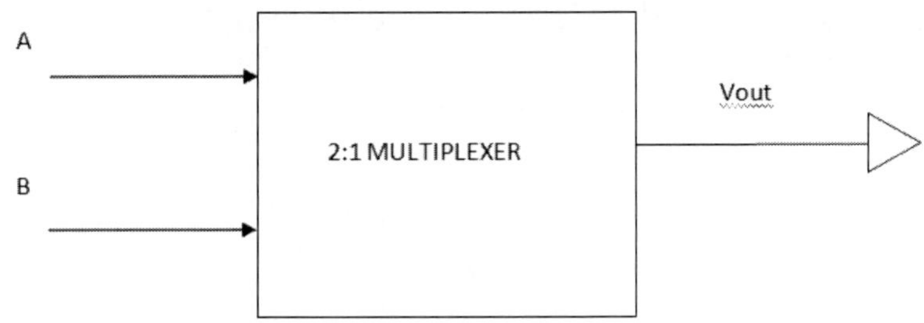

Figure 1.8. Symbol Diagram of 2:1 MUX.

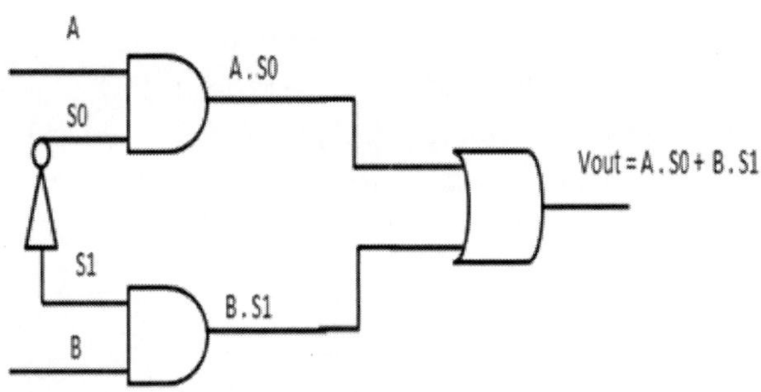

Figure1.9. Gate level Diagram of 2:1 MUX.

1.5.2. Multiplexer 2:1 Integrated Circuit

Now days, serial data communication systems activate at throughputs between 10 and 40 Gb/s. Up till now, communications integrated circuits (ICs) working at such high speeds using GaAs, InP, or SiGe bipolar technologies were engineered. Heavy prominence was placed on to verdict the right match between circuit techniques and fabrication technology. This work demonstrates CMOS to be a practical substitute for broad-band circuit design at 10+ Gb/s. The approach is very inexpensive because of the lower manufacturing costs, higher acquiesce and integration density. Latest achievements in CMOS multiplexer and demultiplexer designs are that the standard CMOS technology are presented which fully develop the speed

potential of 120-nm. Sophisticated circuit techniques and a state-of-the-art fabrication process are mutually to expand speed limits. In high-speed data communication systems, data Multiplexer (MUX) and data Demultiplexer (DEMUX) are input blocks. Present 2:1 MUX previously attained operating speeds of 20+ Gb/s in CMOS [34]-[35]. A 30-Gb/s DEMUX in CMOS has been observed [36] and a complete 10-Gb/s transmitter/receiver with an integrated 16:1 MUX/DEMUX in CMOS has been available [37]. In standard CMOS process with six-layer copper metallization, we have designed a 40-Gb/s 2:1 MUX and 1:2 DEMUX using a 120-nm [38]. The affected Nmos transistors have an f_t of 100 GHz and an clock of 50 GHz [39]. All sub-circuits of the MUX and DEMUX Ics use Current-Mode Logic (CML) with disparity signals. Compared to conventional static CMOS logic, CML allows a decrease of the inner voltage swing. The lower inside voltage swing is vital for high switching speeds. To arrive at this speed, the 2:1 multiplexer uses shunt and series inductive peaking, this virtually doubles the bandwidth. The Block Diagram of 2:1 MUX IC (integrated circuit) is shown in figure 1.11.

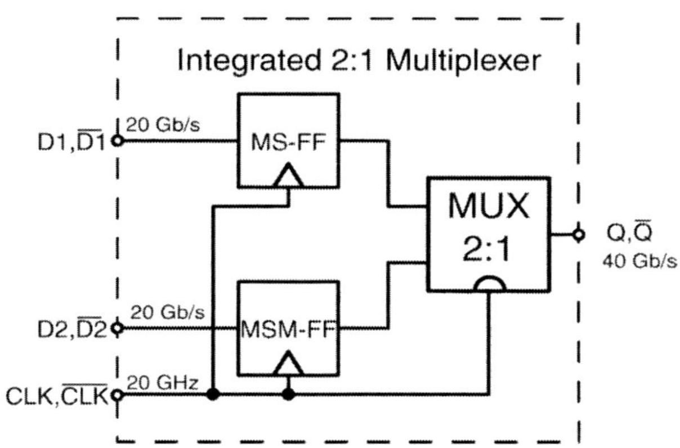

Figure 1.10. Block Diagram of 2:1 MUX IC.

1.6. Logic Styles

1.6.1. Gate Level Multiplexer Design

A multiplexer or MUX is the combinational logic block design that selects the numerous analog or the digital input signal passing ahead to these signals or the selected input signal into the output signal. The multiplexer of 2^n inputs

has a selected line into a simple design of 2:1 multiplexer that is send to the select line to the output. The simple 2:1 MUX is shown with the help of block diagram as shown in figure 1.12.

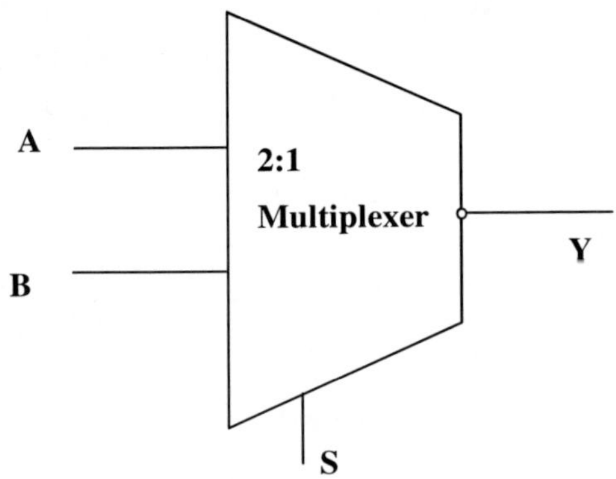

Figure 1.11. Simple Block Diagram of 2:1Multiplexer.

If we design the simple gate level design of multiplexer, the required number of transistors is 20 for the simple operation. The operation for the multiplexer output circuit is given below, Y= S'A+SB

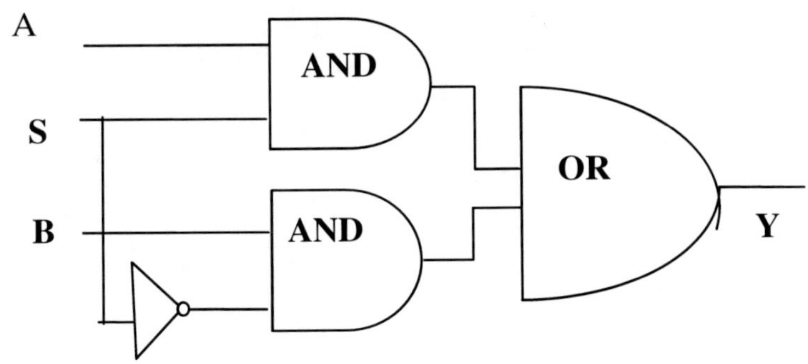

Figure 1.12. Gate Level Design of Multiplexer.

In other two figures 1.13 and 1.14, the design of multiplexer at all levels of gate transistor designs has been shown. In the design of simple gate level or any logic design of VLSI circuits, it requires a large level of the transistor count of the complex design of logic circuit.

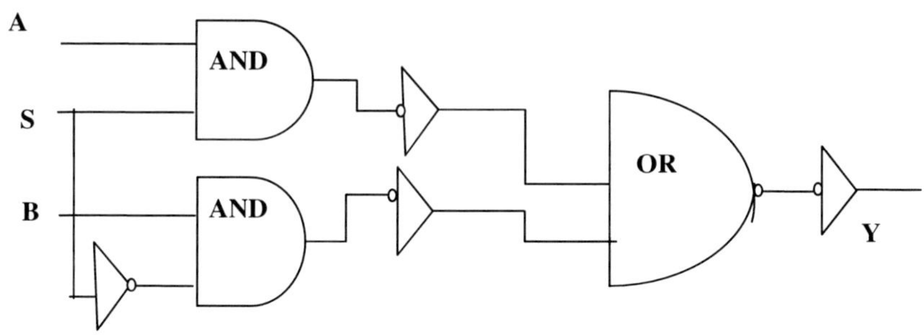

Figure 1.13. Gate Level Mux Design for Levels of Gate.

1.6.2. Design Using Transmission Gate Logic

In VLSI design technology, it is possible to compute the logic function without design of logic gates-the network of MOS transistor can be connected easily together, directly. This is also called the Transmission gates. In this design of transistor we still have gates or to call them differently as logic gates! With the complementary metal oxide semiconductor (CMOS), the NMOS transistors are good conduction of low voltage signal and the PMOS is good conduction of high voltage level signals. So, the design of transmission gate is using the pair of complementary transistors. When the control signal S is possible to high, the transmission gate conducts the logic signal voltage of either sense or not in sense of direction. The special symbol is used for transmission gate [40]. Denoted in figure 1.15. and 1.16.

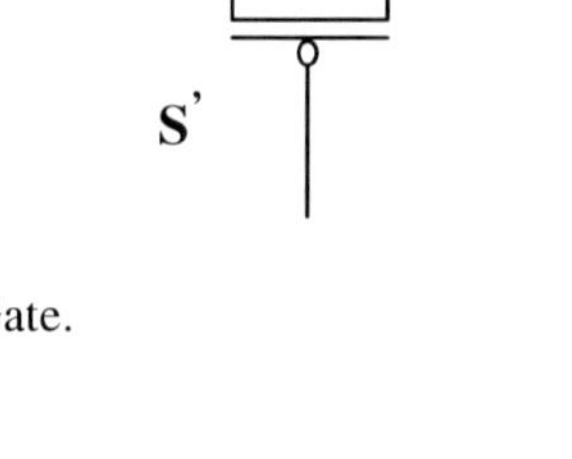

Figure 1.14. Transmission Gate.

Figure 1.16 presents forth design of transmission gate multiplexer. This multiplexer design uses the transmission gate based multiplexer and generates their block diagram of 2:1 multiplexer, that the two inputs A, B and S select input, the output Y depends on the select line into the multiplexer design.

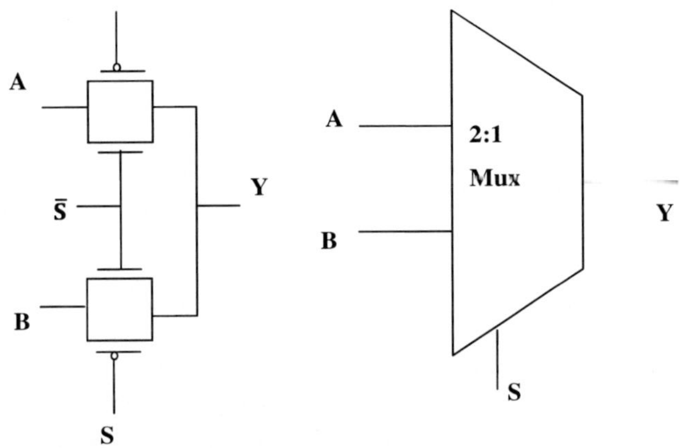

Figure 1.15. Transmission gate Multiplexer.

1.6.3. Design Using Pass Transistor Logic

The MOS transistor can be used as a simple switch to steer the current using the pass transistor logic. Let's take a simple 2:1 multiplexer with the two control inputs S and S' four data inputs and the output signal Y. The function can be easily specified by a simple logic. Unfortunately, the power consumption will be problem here because the input to the inverter will be at the Vdd-Vt. This means that the PMOS is not fully turned 'off' in this condition. This problem gets worse as you get deeper into advance Nanoscale technology. As shown in figure 1.17, this concept of pass gate transistor level requires the small number of transistor count in comparison to the other gate level design of logic circuit. This is the basic design concept of recent trends in VLSI technology scaling and one of the easiest and the most complex design circuitry without degradation in the circuit. This design methodology is used more preciously for the reversible logic gate.

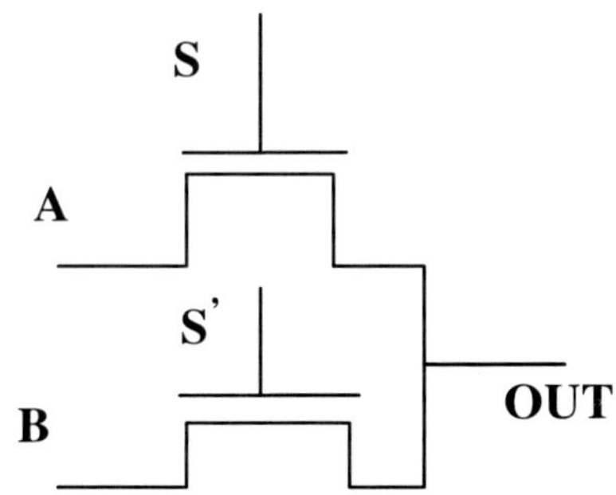

Figure 1.16. Simple 2:1 multiplexer.

1.7. Impact of Logic Style

The logic style used in logic gates basically influences the speed, size, power dissipation and the wiring complexity of a circuit. The circuit delay is determined by the number of inversion levels, the number of transistors in series, transistor sizes (i.e., channel widths) and intra- and intercell wiring capacitances. Circuit *size* depends on the number of transistors and their sizes and on the wiring complexity.

1.7.1. Logic Style Requirements for Low Power

According to the formula,

$$Pdyn = V^2dd.fclk \sum_n \alpha n.cn + Vdd.\sum_n iscn$$

The dynamic power dissipation of a digital CMOS circuit depends on the supply voltage Vdd, the clock frequency fclk, the node switching activities αn, the node capacitances cn, the node short-circuit currents iscn, and the number of nodes n all the other parameters are influenced, to some degree, by the logic style applied. Thus, some general logic style requirements for low-power circuit implementation can be stated at this point.

a) Switched Capacitance Reduction:

Capacitive load, originating from transistor capacitances (gate and diffusion) and linked wiring, is to be minimized. This is achieved by having the minority transistors and circuit nodes as achievable, and by dipping transistor sizes to a minimum. An additional source for capacitance reduction is established at the layout level [41].

b) Supply Voltage Reduction:

The choice of logic style and the supply voltage are ultimately related through delay-driven voltage scaling. A logic style must be vigorous against supply voltage diminution, i.e., performance and correct functioning of gates must be certain at low voltages as well. This becomes a rigorous problem at very low voltages of around 1 V and lower, where noise margins become vital [42]-[43].

c) Switching Activity Reduction:

Switching activity of a circuit is primarily controlled at the architectural and Registers Transfer Level (RTL). At the circuit level, large differences are mainly practical between static and dynamic logic styles.

d) Short-Circuit Current Reduction:

Short-circuit currents (also called dynamic leakage currents or overlie currents) may diverge by a substantial amount between different logic styles. They also sturdily depend on input signal slopes (i.e., vertical and balanced signal slopes are superior) and thus on transistor sizing.

1.7.2. Logic Style Requirements for Ease-of-Use

For ease-of-use and generality of gates, a logic style should be highly robust and have friendly electrical characteristics, that is, decoupling of gate inputs and outputs (i.e., at least one inverter stage per gate) as well as good driving capabilities and full signal swings at the gate outputs, so that the logic gates can be cascaded arbitrarily and work reliably in any circuit configuration. These properties are prerequisites for cell-based design and logic synthesis, and they also allow for efficient gate modeling and gate-level simulation.

1.7.2.1. Reversibility of Physics and Reversible Computing

In the implementation of reversible logic computing, it is essential to use the learning to characterize and control the physical dynamics mechanisms to carry the reversible logic gate computational operation. We can accumulate this to negligible total amount of the uncertainty with regards to the complete physical state operation of these mechanisms, for each and every operation performed by the reversible logic operation. In other form, we can say that the precise track of the state of active energy is involved to carry out the computational operation inside the machine, and the majority of this energy is recovered in this organization that can be used or reused for subsequent operation, rather than permitted to dissipate in the form of heat.

$$\text{Solution} = \text{Reversibility} \tag{1.2}$$

The field of research awaits the development of high quality, cost effective and near to the reversible logic device advance down scaling technology. One of the most important elements is that it includes the highly energy efficient logic clocking rate and synchronization mechanisms. This solid state engineering progress will be the mostly needed for future prospects before the large computing theoretical research on reversible operation application.

Reversible logic circuits were first developed in the year of 1960s by the theoretical consideration of zero-energy computation of reversible logic as well as the practical improvement for the bit manipulation transformation in the cryptography and computer graphics. Since the year 1980s, 'reversible circuits' have attracted more interest as components of the quantum computer algorithm and it is more recently in photonic and Nanoscale-computing technologies where some switching device offer no signal gain of the reversible logic [44]. The surveys of reversible logic circuits are becoming very important [45]. Their construction and optimization as well as the recent research challenges are available now days for reversible logical and physical operation [46].

1.7.2.2. Reversible Logic/Reversibility

Reversible logic is a very prospective approach of logic synthesis for power reduction in future computing technologies [47]. In a reversible logic circuit, the number of inputs and outputs are same and there is a one-to-one mapping between input and output values. Reversible circuits are constructed using reversible gates. The recent computers erase a bit of information every

time they perform a logical operation. Such logical operations are generally known as ‘irreversible logic’. They impose many design constraints. Zero energy dissipation would be possible only if the network consists of reversible gates. So that the reversibility will become an essential property in future circuit design [48]. According to Landauer, traditional irreversible hardware computation inevitably leads to energy dissipation due to the loss of each one bit of information which dissipates an amount of KTln (2) Joules of energy, where K is the Boltzmann’s constant (1.3807×10^{-23} JK^{-1}) and T is the absolute temperature at which computation is performed [49]. In the case of high technology circuits, the heat dissipation is very large, when compared to a loss of one bit information at room temperature. Furthermore, voltage-coded logic signals have energy of $E=1/2CV^2$ and this energy is dissipated whenever the node voltage changes in the irreversible CMOS technology. It is estimated that reversible logic circuits also help to secure energy by using charge recovery logic [50]. As the Moore’s law continues to hold, the processing power just doubles every 18 months. The present irreversible technologies can dissipate a lot of heat and can reduce the life of the circuit. The reversible logic operations do not lose (erase) information and dissipate very less heat. Thus, reversible logic circuits are likely to be in demand in high-speed power aware circuits. Reversible logic circuits are of high interest in low-power CMOS design, nanotechnology, quantum computing and optical computing. The most important application of reversible logic lies in quantum computers [51]. A quantum computer will be viewed as a quantum network (or a family of quantum networks) composed of quantum logic gates; each gate performing an elementary unitary operation on one, two or more than two-state quantum systems known as Qubits. The gate/circuit that does not loose information or data is called reversible logic gate. A reversible logic circuit has the facility to generate a unique output vector from each input vector and vice versa.

A reversible logic gate is n × k logic device where n and k are the number of inputs and outputs with one-to-one mapping and should be equal to k (n = k). A gate is said to be reversible if the input vector can be uniquely recovered from the output vector and there is a one-to- one correspondence between its input and output logic [52]. A set of reversible logic gates is needed to design reversible logic circuits. An n × k reversible logic gate can be represented as:

Input Vector

$$I_V = (I_{i,j}, I_{i+2,j}, I_{i+3,j} \dots \dots \dots\ I_{k-1}, I_{k,j}) \qquad (1.3)$$

Output Vector

$$O_V = (O_{i,j}, O_{i+2,j}, O_{i+3,j} \dots \dots \dots O_{k-1,j,} O_{k,j}) \quad (1.4)$$

For each particular Vector $I_V = O_V$

Reversible are circuits in which the number of inputs is equal to the number of outputs and there is one-to-one mapping between vectors of input and outputs as shown in figure1.18 given below.

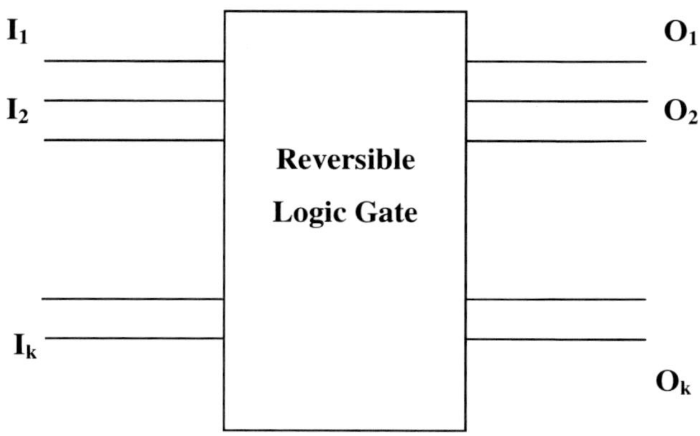

Figure 1.17. Gate with K Input and Output is called K*K Gate.

The reversible logic is a more alternative and traditional approach for irreversible logic gate circuits. Since the reversible logic computing does not erase or lose their information or data in the logical operation, as a result of reversible logic having a theoretical potential to dissipate no amount of energy. According to the Frank [53], reversible logic can easily recover a fraction of energy i.e. reach up to the level of 100%. As there is no limit to reduce the dissipation of heat in reversible logic circuits, the dissipated amount of the heat will became very close to zero with the development of hardware.

Reversible Logic play a vital role in VLSI design technology of quantum computing since the quantum computation is the reversible process, and the most physical reality in this quantum logic operation can be easily demonstrated [54]. Interested readers can easily refer to the detailed discussion of the Quantum computing network [55]. The reversible computing is also the most useful for the other Nanoscale technologies including the low power, low leakage current and better frequency response for CMOS design [56], Optical computing [57], nanotechnology and the digital communication [58].

1.7.3. Difference between Reversible and Irreversible Gate

Losing any information in the logic gate circuit is causing losses to its power performance. When the loss of information into the input vector cannot be uniquely recovered in the form of output vector of combinational logic gate operation, this operation is said to be a reversible. The recent computers erase a bit of information every time they perform a logical operation. Such logical operations are generally known as 'Irreversible logic', as shown in truth table 1.2 for the two input A, B, and the output P=A, Q=AΘB.

Table 1.2. Truth Table for Reversible EXOR gate logic

Input		Output	
A	B	P=A	Q=AΘB
0	0	0	0
0	1	0	1
1	0	1	1
1	1	1	0

1.7.4. Reversible Logic Gate

The simplest reversible gate is NOT gate and is a 1×1 gate. The controlled NOT (CNOT) gate is an example of a 2×2 gate. Many of the available gates are 3×3 Reversible gates such as TR, FG, PG and TG gates. The Quantum Cost of 1×1 Reversible gates is zero, and the Quantum Cost of 2×2 Reversible gates is one. Any reversible gate is optimized by using 1×1 NOT gates (inverter) and 2×2 (Feynman gate) Reversible gates, such as V, V^+ (V is square root of NOT gate and V^+ is its Hermitian) and FG gate also known as CNOT gate. The properties of V and V^+ Quantum Gates are given in the following equations [59].

$$V \times V = NOT \quad (1.5)$$

$$V \times V^+ = V^+ \times V = I \quad (1.6)$$

$$V^+ \times V^+ = NOT \quad (1.7)$$

The Quantum Cost of a Reversible gate is calculated by counting the number of V, V^+ and CNOT gates. The two gates i.e. NOT gate and CNOT gate (Controlled NOT gate) are shown in figure 1.4.

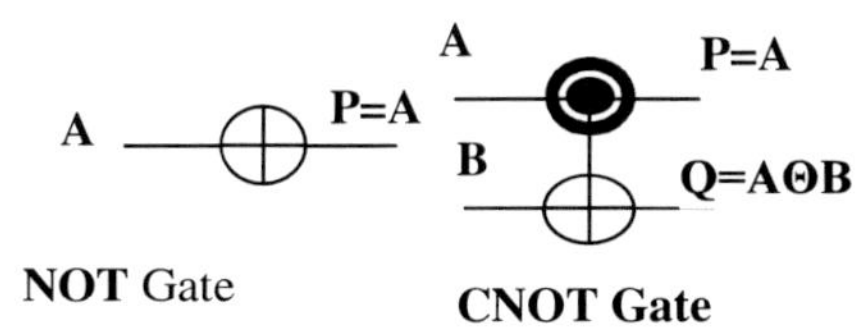

Figure 1.18. Combinational Circuit Diagram of 1x1 NOT Gate and CNOT Gate.

(i) *Garbage Input/Output*-One of the most important features of a reverse gate is its garbage output, i.e. each and all input of the gate that is not used as input to another gate or as a primary output is called garbage output. The unutilized outputs from a gate arc called 'garbage'. Heavy price is paid off for every garbage output [60]. Let's want to determine the Ex-OR between the two variables in the reversible logic computing, then this circuit will look like figure 1.20. The one extra output of this operation should be produced to make the logic circuit reversible and that's unwanted output (P = A) also known as the garbage.

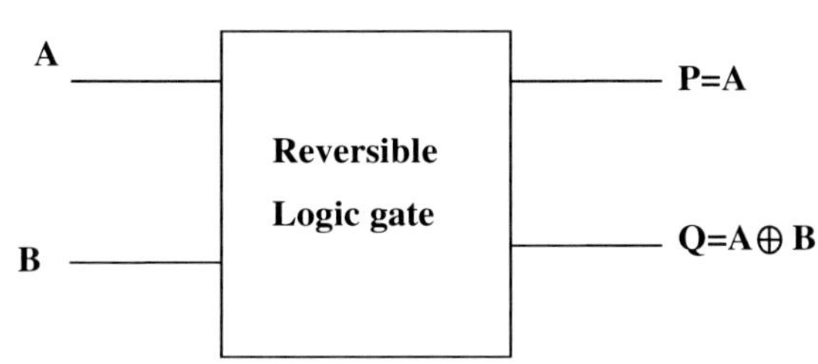

Figure 1.19. Basic Block of Reversible Logic Gate with Garbage Input/Output.

(ii) Quantum Cost-These constants are used as the input line that is either set to Zero (0) or One (1) in the circuit input side. Gate Count is the number of reversible gates used to realize the function. The Hardware Complexity refers to the number of basic gates (NOT, AND, and EXOR gate) used, easy to synthesize given more functions [61].

(iii) Critical Path-The delay can be calculated in reversible logic based on the critical path. It is the longest path that gives the desired output of the system.
(iv) Gate Count- Any reversible circuit realizes only the function that is reversible.
(v) Transistor Count-The transistor count of the reversible gate is to realize effort to need the number of transistors used in the logical gate [62].

There exist many reversible gates in the literature. Among them Feynman gate, (FG) Peres gate (PG), Toffoli gate (TG), Fredkin gate (FRG) and Khan Gate (NG) are mostly common used [63]. The quantum realizations of all these gates are not available in the literature. Only PG, FG, TG and FRG have been realized in nanotechnology [64].

1.8. Proposed Reversible Multiplexer

The proposed design of reversible multiplexer design requires the less number of transistors and the design implementation of the complex design of large logic gate circuitry. This is the most fundamental approach for every design of logic gate is the using pass transistor logic gate concept.

1.8.1. 2-to-1 Reversible Multiplexer Using Pass Transistor

The design of 2:1 multiplexer using the number of transistor count is less as compared to the other gate level design of simple multiplexer circuit. In other words, we can say this is less area and less number of transistor count with small amount of delay.

1.8.2. Basic Concept of Design 2:1 Reversible Multiplexer Function

The proposed reversible 2:1 multiplexer gate is realized using transistor implementation as described in Figure 1.21. The required output Y can be obtained using only two transistors. However, three transistors are required for calculating garbage outputs. To obtain output G1, a pass transistor is used for

passing the selection input S (select) to the output G1 as shown in Fig 1.13. (c). In Fig1.13, when the selected input is 0, the PMOS transistor will be ON while NMOS will be OFF, thus input A is passed to the output Y. If selection input is 1, then the NMOS transistor will be conducting while PMOS is in OFF.

State passing input B to the output Y. Similar to output Y, output G2 is passed with input B when S=0 (SEL) and input A, when S=1 [65].

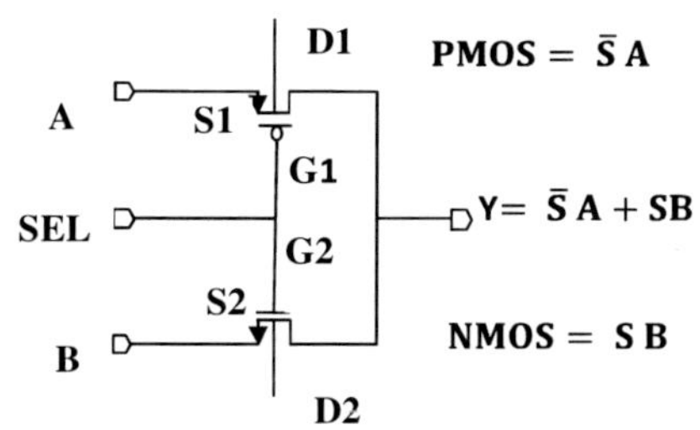

(a) Circuit Diagram for Multiplexed Output Y

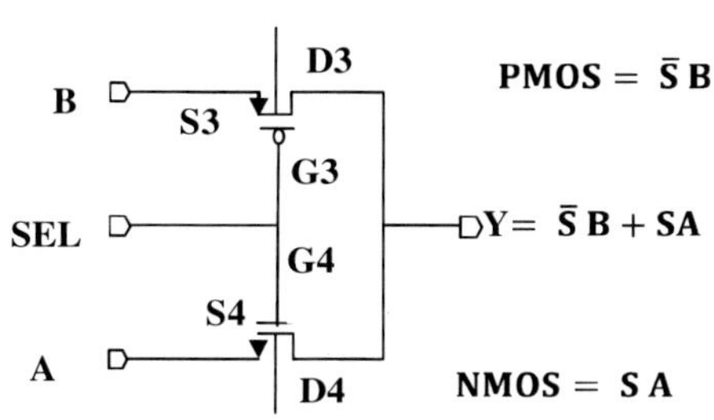

(b) Circuit diagram for output G2

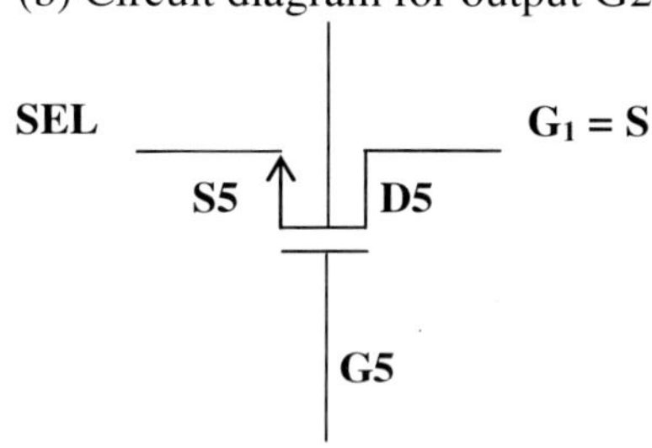

(c) Circuit diagram for output G1.

Figure 1.20. Implementation of Reversible Multiplexer with Garbage Input/ Output Gate (a) Circuit Diagram for Multiplexed Output Y (b) Circuit Diagram for Output G2 (c) Circuit Diagram for Output G1.

REFERENCES

[1] G. Busch, "Early history of the physics and chemistry of semiconductors-from doubts to fact in a hundred years," *European Journal of Physics,* vol. 10, no. 4, pp. 254–264, 1989.

[2] F. Laeri, F. Schüth, U. Simon, and M. Wark, *"Host-Guest-Systems Based on Nan porous Crystals,"* Eduardo Ruiz-Hitzky, Wiley-Verlag GmbH & Co. online library Weinheim, pp. 435–436, 2003.

[3] F. Braun, "On the current conduction in metal sulphides," in Sze, S.M. *Semiconductor Devices: World Scientific Publishing Co. Singapore,* vol. 153, no. 4, pp. 377-380, 1991.

[4] Schuster, A., "On Unilateral Conductivity," *Philosophical Magazine, Taylor & Francis online,* vol. 48, no.318, pp. 251-258, May 2009.

[5] L. Hoddeson, E. Braun, J. Teichmann, and S. Weart, "Out of the Crystal Maze: Chapters in the History of Solid State Physics," *Critical Assembly: A Technical history of Los Alamos during the Oppenheimer,* New York: Oxford University Press, pp.728, 1992.

[6] Moreau, R. "The Computer Comes of Age – The People, the Hardware, and the Software," London: The MIT Press, *The British Journal for the history of Science,* vol.18, no.3 pp.372-373, 1985.

[7] Rojas, Raul and Ulf Hashagen. *"The First Computers – History and Architecture,"* London: The MIT Press, Thomas Publisher, San Francisco, pp.131-153, 2000.

[8] S. McCartney, *"ENIAC: The Triumphs & Tragedies of the World's First Computer,"* Walker & Company New York, pp.124-127, 1999.

[9] B. Lojek, *"History of Semiconductor Engineering,"* first Edition, pp.3-9, July 2010.

[10] J. Orton, *"the Story of Semiconductors,"* Oxford University Press, pp.260-267, 2004.

[11] R. Riedlinger Arnold, R. Biro, L. Bowhill, B. Crop, J. Duda, K. Fetzer, E.S. Franza, O. Grutkowski, "A 32 nm, 3.1 billion transistor, 12 wide issue Itanium processor for mission-critical servers," *IEEE Journal of Solid-State Circuits,* vol. 47, no. 1, pp.177-193, January 2012.

[12] M. Bohr and K. Mistry, "Intel's revolutionary 22 nm transistor technology," *Intel Corporation,* pp.1-27, May 2011.

[13] F. M. Wanlass and C. T. Sah, "Nanowatt Logic Using Field-Effect Metal-Oxide Semiconductor Triodes," *Proceeding of IEEE International Conference on Solid State Circuits,* pp. 32-33, February 1963.

[14] T. S. Perry, Gordon Moore's Next Act," *Journal IEEE Spectrum,* vol. 45, no. 5, pp.38-47, May 2008.

[15] By Lanny L. Lewyn, Life Senior Member IEEE, Trond Ytterdal, Senior Member IEEE, Carsten Wulff, Member IEEE, and Kenneth Martin, Fellow IEEE "Analog Circuit Design in Nanoscale CMOS Technologies," *Proceeding of IEEE Journal,* pp. 1687-1714, vol. 97, no. 10, October 2009.

[16] *The International Technology Roadmap for Semiconductors* 2004 (ITRS), pp.1-13, July 2004.

[17] Rajeev R. Rao, Anirudh Devgan, David Blaauw, "Parametric yield estimation considering leakage variability," *Proceedings of IEEE 41st Annual Design Automation Conference (DAC'04),* New York, USA, pp.442-447, 2004.

[18] Mohab Anis and Mohamed H. Aburahma "Leakage Current Variability in Nanometer Technologies," *Proceedings of IEEE 5th International Workshop on System-on-Chip (SOC) for Real-Time Application,* pp.60-63, 2005.

[19] S. Borkar, Tanay Kanik, Shiva Narendra, Jim Tschanz, "Parameter Variations and Impact on Circuits and Micro architecture," *Proceedings of IEEE 40th Annual Design Automation Conference (*DAC 2003), pp. 338-342, 2003.

[20] R. H. Dennard, F. H. Gaensslen, H. N. Yu, V. L. Rideout, E. Bassous, and A. R. LeBlanc, "Design of ion-implanted MOSFET's with very small physical dimensions," *IEEE Journal of Solid-State Circuits Newsletter,* vol. 12, no.1, pp. 38-50,2007.

[21] M. Campbell, V. Gerousis, J. Hogan, J. Kibarian, L. Lanza, W. Ng, D. Pramanik, A. Strojwas, and M. Templeton, "When IC yield missed the target, who is at fault?," *Proceeding of IEEE 41st Design Automation Conference (ACE DAC'04)* pp. 80, Jun. 2004.

[22] B. Razavi, "Design of Analog CMOS Integrated Circuits," New York: Mc-Graw-Hill, pp.212-249, 2001.

[23] K. Kim, T. Ahn, S.-Y. Han, C.-S. Kim, and K.-H. Kim, Low power multiplexer decomposition by suppressing propagation of signal transitions, *Int'l Symp. on Circuits and Systems*, vol. 5, pp. 85-88, 2001.

[24] Sebastian T. Ventrone, Low power multiplexer circuit, *United States Patent*, Patent Number: 6,054,877, Apr. 2000.

[25] P. Metzgen "A High Performance 32-bit ALU for Programmable Logic", *Proceedings of the 2004 ACM/SIGDA 12th international symposium on Field Programmable Gate Arrays*. Pp 61-70. 2004.

[26] *D.* Chen and J. Cong, "*Register Binding and Port Assignment for Multiplexer Optimization*", ASPDAC, Jan. 2004.

[27] D. Chen and al, "*Low-Power High-Level Synthesis for FPGA Architectures*" Int. Symp. Low Power Elec. and Design, Aug. 2003.

[28] K. Kim and al, "Low power multiplexer decomposition by suppressing propagation of signal transitions", *Int'lSymp. on Circuits and Systems,* vol. 5, pp. 85-88, 2001

[29] Sebastian T. Ventrone, "*Low power multiplexer circuit*", United States Patent, Patent Number: 6,054,877, Apr. 2000

[30] K. Jabeur, D. Navarro, I. O'Connor, P.E. Gaillardon, M.H. Ben Jamaa, F. Clermidy, "*Reducing transistor count in clocked standard cells with ambipolar double-gate FETs*", IEEE/ACM international symposium on nanoscale architectures (Nanoarch'10), june 17-18 2010, Anaheim, CA, USA.

[31] Kotb Jabeur, Natalya Yakymets, Ian O'Connor and Sébastien Le Beux. "*Fine-Grain Reconfigurable Logic Cells Based on Double-gate CNTFETs*", IEEE/ACM International Conference on Great Lakes Symposium on VLSI (GLSVLSI), Lausanne, June, 2011

[32] Kotb Jabeur, Natalya Yakymets, Ian O'Connor and Sébastien Le Beux. "*Ambipolar double-gate FET binary-decision- diagram (Am-BDD) for reconfigurable logic cells*", International Symposium on Nanoscale Architectures (NANOARCH), San Diego, June, 2011.

[33] F. Moradi, D. Wisland, H. Mahmoodi, Y. Berg and T. Cao, "New SRAM Design Using Body Bias technique for ultra low power application," *11th Int'l Symposium on Quality Electronic Design,* pp. 468-471, 2010.

[34] Daniel Kehrer, Hans-Dieter Wohlmuth, Herbert Knapp, Martin Wurzer and Arpad L. Scholtz, "40-Gb/s 2:1 Multiplexer and 1:2 Demultiplexer in 120-nm Standard CMOS*", IEEE Journal of Solid-State Circuits,* vol. 38, NO. 11, November 2003.

[35] T. Suzuki, Y. Nakasha, T. Sakoda, K. Sawada, T. Takahashi, K. Makiyama, T. Hirose, and M. Takigawa, "A lOO-Gbit/s 2: 1 Multiplexer in InP HEMT Technology", IEEE MTFS Digest year 2003.

[36] Yoshiyasu Doi, Syunitirou Masaki, Takaya Chiba, Hirohito Higashi, Hisakatsu Yamaguchi, Hideki Takauchi, Hideki Ishida, Kohtaroh Gotoh, Junji Ogawa, Hirotaka Tamura, "A 0.8-1.3V 16-channel 2.5Gb/s High-speed Serial Transceiver in a 90nm Standard CMOS Process", *IEEE 2 Custom Integrated Circuits Conference*, year 2005.

[37] Sara Hashemi, Mostafa Rahimi Azghadi, Ali Zakerolhosseini, *"A Novel QCA Multiplexer Design"* Internatioal Symposium on Telecommunications year 2008.

[38] Filipe G. R. Ramos, *"A Multi-Voltage Reference Source",* Fourth Congress of Electronics, Robotics and Automotive Mechanics, year 2007.

[39] Ahmad Yazdi and Michael M. Green, "A 40-Gb/s Full-Rate 2:1 MUX in 0.18-um CMOS" *IEEE Transactions on Microwave Theory and Techniques,* vol. 59, NO. 11, November 2011.

[40] VLSI design *"University of Cambridge Computer Laboratory,"* New Museums Site Pembroke Street Cambridge CB2 3QG, pp.1-25, 1974.

[41] C. Piguet, J.-M. Masgonty, P. Mosch, C. Arm, and V. von Kaenel, "Low-power low-voltage standard cell libraries," in *Proc. Low Voltage–Low Power Workshop,* ESSCIRC'95, Lille, France, Sept. 1995.

[42] C. Piguet, J.-M. Masgonty, S. Cserveny, and E. Dijkstra, *"Lowpower low-voltage digital CMOS cell design,"* in Proc. PATMOS'94, Barcelona, Spain, Oct. 1994, pp. 132–139.

[43] N. Ohkubo et al., "A 4.4 ns CMOS 54 _ 54-b multiplier using pass-transistor multiplexer," *IEEE J. Solid-State Circuits,* vol. 30, pp. 251–257, Mar. 1995.

[44] Rolf Drechsler, Robert Wille. "From Truth Tables to Programming Languages: Progress in the Design of Reversible Circuits," *Proceedings of the 41st IEEE International Symposium on Multiple-Valued Logic (ISMVL),* IEEE Computer Society Washington, DC, USA, pp. 78-85, 2011.

[45] Mehdi Saeedi, Igor L. Markov, *"Synthesis and Optimization of Reversible Circuits,"ACM Journal on Emerging Technology in computing System (JETC),* Computing Surveys (CSUR) ACM New York, USA, vol. 45, no. 2, pp.1-34, February 2013.

[46] Rolf Drechsler and Robert Wille, *"Reversible Circuits: Recent Accomplishments & Future Challenges for an Emerging Technology,"* Proceeding of 16th International Symposium on VLSI Design and Test (VDAT), India, pp.383-392, July 2012.

[47] A. De Vos. Reversible *"Computing: Fundamentals, Quantum Computing, and Applications,"* Wiley-VCH Verlag GmbH & Co. (Online Library) KGaA, Weinheim, Germany, pp.261, October 2010.

[48] S. N. Mahammad and K. Veezhinathan, "Constructing online testable circuits using reversible logic," *IEEE Transactions on Instrumentation and Measurement,* vol. 59, no.1, pp.101–109, 2010.

[49] D. Maslov. *"Reversible logic synthesis,"* PhD thesis, University of New Brunswick, pp.11-19, 2003.

[50] R. C. Merkle and K. E. Drexler. *"Helical logic,"* Published in Nanotechnology vol.7, no 4, pp.325–339, May 1996.

[51] N. M. Nayeem and J. E. Rice. "Online fault detection in reversible logic," *Proceedings of IEEE International Symposium on Defect and Fault Tolerance in VLSI and Nanotechnology Systems (DFT),* Vancouver, BC, Canada, pp.426-434, October 2011.

[52] Irina Hashmi and Hafiz Md. Hasan Babu, "An Efficient Design of a Reversible Barrel Shifter," *Proceeding of Twenty Third International Conference on VLSI (VLSID '10),* Design, pp. 93-98, Jan. 2010.

[53] M. P. Frank. "Introduction to reversible computing: motivation, progress, and challenges," *Proceedings of the 2nd Conference on Computing Frontiers, Italy and New York, USA,* pp. 385–390, May 2005.

[54] A. N. Al-Rabadi. *"Reversible Logic Synthesis: From Fundamentals to Quantum Computing, Electronics & Electrical Engineering Computing,"* Springer-Verlag, Springer Berlin Heidelberg, pp.186-204, 2004.

[55] M. Nielsen and I. Chuang. *"Quantum Computation and Quantum Information,"* First edition, Cambridge University Press, pp.676-700, 2000.

[56] W. C. Athas and L. J. Svensson. "Reversible logic issues in adiabatic CMOS," *Proceedings of Workshop on Physics and Computation (Phys Comp) Dallas,* TX, USA, pp. 111-118, 1994.

[57] H. Rahaman, D. K. Kole, D. K. Das, and B. B. Bhattacharya. "On the detection of missing-gate faults in reversible circuits by a universal test set," *Proceedings of the 21st International Conference on VLSI Design,* pp.163–168, 2008.

[58] R. C. Merkle. *"Reversible electronic logic uses switches,"* Published in Nanotechnology, vol.4, no.1, pp.21–40, 1993.

[59] M. S. Islam and M.R. Islam, "Minimization of Reversible Adder Circuits," *Asian Journal of Information Technology,* vol.4, no.12, pp. 1146-1151, 2005.

[60] M. Haghparast, S. Jassbi, K. Navi, and O. Hashemipour, "Design of a novel reversible multiplier circuit using HNG gate in nanotechnology," *Journal of World Applied Sciences,* vol. 3, no. 6, pp. 974–978, 2008.

[61] N.M. Nayeem, M.A. Hossain, L. Jamal, H.M.H. Babu, "Efficient Design of Shift Registers Using Reversible Logic," *Proceeding of IEEE*

International Conference on Signal Processing Systems, pp.474-478, May 2009.

[62] S. Gorgin, A. Kaivani, "Reversible Barrel Shifters", *ACS/ Proceeding of IEEE International Conference on Computer Systems and Applications (AICCSA'07)*, pp. 479-483, May 2007.

[63] H.R. Bhagyalaksmi, M.K. Venkatesha, "Optimized reversible BCD adder using new reversible logic gates," *Journal of Computing,* vol. 2, no.2, pp.28-32, February 2010.

[64] S. Kotiyal, H. Thapliyal, N. Ranganathan, "Design of A ternary barrel shifter using multiple-valued reversible logic," *Proceeding of IEEE 10th Conference Nanotechnology (IEEE-NANO),* pp.1104- 1108, Aug. 2010.

[65] Praveen.B, Vinay Kumar S.B "2^n:1 Reversible Multiplexer and its Transistor Implementation," *International Journal of Engineering Science and Technology (IJCER),* vol. 2, no.1, pp.182-189, Jan-Feb 2012.

Chapter 2

Impact of Techniques in Nanosclae CMOS Technology

2.1. Objectives of the Chapter

Today's focus on low power is not the only reason for that of the current growing demands of mobile applications. Even prior to the mobile era, power consumption has been an essential problem. To resolve the power consumption problem, many researchers have proposed diverse thoughts from the device level to the architectural level and above. There is no universal technique to avoid tradeoffs between power, delay and area, thus designers are requisite to choose suitable techniques that gratify application and product requirements. With the recent enhanced CMOS technology scaling for reversible logic computing capability, the feature size of integrated circuit miniatures the Transistor in the reversible logic gate low power with garbage input/output for a design circuit that is 2:1 reversible multiplexer. Presented here is the functionality of the circuits frequency response of the logic gate operation of 2:1 reversible multiplexer which is increasing the integrated circuits with each new generated technology by the enhancing the future technology. This chapter also describes the various leakage reduction techniques in CMOS Multiplexer design.

2.2. Technology Scaling and Low Power Circuits

Power consumption was a section of the design process, but was less perceptible. Still, with the attainment structures below 0.18μm and having high-performance-chips exertion in portable devices, power dissipation has become the key design unease in these applications. This addresses designers and provides as instruction for system level specification and the remains of a Low Power Design Flow [1]. Depending on the application, the kind of circuit to be realized and the design technique used, diverse recital aspects become significant, disallowing the formulation of universal convention for best possible logic styles [2]. As driven by Moore's law, VLSl systems are becoming progressively more complex [3], and the advances in Deep Sub-Micron (DSM) technologies have made probable multi-million gate designs, so the circuit designers are facing a incredible challenge of interconnect-centric design flow.

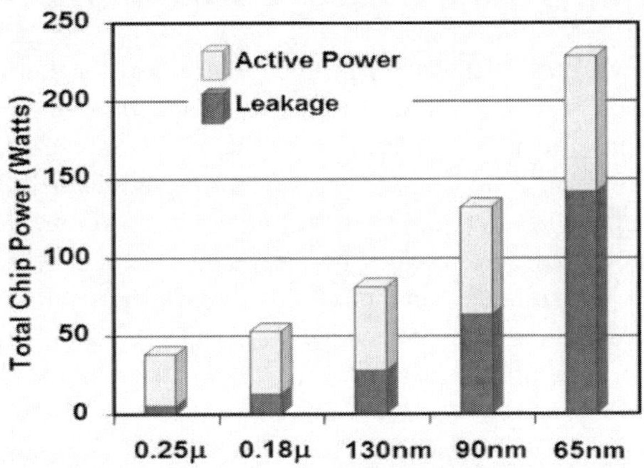

Figure 2.1. Impact of Technology on Total Chip Power.

Power consumption is one of the peak concerns of Very Large Scale Integration (VLSI) circuit design, for which Complementary Metal Oxide Semiconductor (CMOS) is the principal technology. Today's focus on low power is not only for the reason of the current growing demands of mobile applications. Even prior to the mobile era, power consumption has been an essential problem. To resolve the power consumption problem, many researchers have proposed diverse thoughts from the device level to the architectural level and over. Though, there is no universal technique to avoid

tradeoffs between power, delay and area, and thus designers have been requisite to choose suitable techniques that gratify application and product requirements. Power dissipation of CMOS consists of dynamic and static components. When transistors are switching, dynamic power and static power are dissipated in spite of transistor switching. Dynamic power dissipation was earlier (at 0.18μ technology and above) the single major concern for low-power chip designers since dynamic power accounted for 90% or more of the total chip power. Consequently, many earlier proposed techniques were formulated, such as voltage and frequency scaling, paying attention on dynamic power diminution. Conversely, as the characteristic size shrinks, e.g., to 0.09μ and 0.065μ, static power has resided into an immense challenge for existing and prospect technologies. It depends on the International Technology Roadmap for Semiconductors (ITRS) [4], Kim et al. report that sub-threshold leakage power dissipation of a chip may top dynamic power consumption at the 65nm feature size [5]. Many reserves in power consumption can be obtained by combining different techniques at diverse levels of design abstraction from RTL (Register Transfer Level) [6-7] to transistor level. To reduce average power consumption in gate level circuits we focus on the technique, especially in multiplexer hierarchy. With the rising use of portable consumer electronics, power consumption has become a considerable performance trait for a chip due to both partial battery life in portable systems and also due to exclusive packages and heat sinks necessary for high power levels. Accordingly, the design of low-power digital systems is becoming increasingly significant.

2.3. Low Power Digital CMOS Design

The recently enhanced CMOS technology scaling for reversible logic computing capability of the integrated circuits [8] which increased the feature size of integrated circuit miniaturized the transistor in this reversible logic gate low power design circuit i.e. 2:1 reversible multiplexer with garbage input/output [9], thereby the enhancing the future technology and the functionality of the circuits [10]. Frequency response of the logic gate operation of 2:1 reversible multiplexer is increasing the integrated circuits with each new generated technology [11]. The lower power consumption in the integrated circuits is increasing with the higher number of transistors and enhancing their frequency response and the periodic jitter of these reversible operational circuits. To most power consumption in reversible logic has

become one of the primary hindrances into the further research into the advanced Nano scale CMOS technology integrated circuits [12]. The low power reversible logic design is an important research area in recent trends for advance technology. There are the three basic and primary issues for the low power stimuli circuit design. One is the environment friendly green computing battery life is easier for the powerful mobile application and the emerging life of the self-sustaining/battery replacement-for free application such as the intelligent network sensing nodes.

2.3.1. Gate Power Dissipation in CMOS

The dreading continues for the feature size of the transistor which correspondingly increases the density of chip with every new generation of the technology. The operating frequencies for the reversible logic gate have made power consumption a major issue in the recent VLSI design technology. Most of the power dissipation in the reversible logic circuit in the ICs discourages their use in the portable systems. It also is the region of overheating in reversible logic gate, which is decreasing the performance and reducing the area of the chip for life time. To control the temperature level of the reversible logic, designed chip have needed special cost of packaging and cooling system of the reversible logic arrangements, that provide the appropriate result in the future escalation of the semiconducting device.

2.3.2. Power Dissipation in CMOS Circuits

There are the three main sources of power dissipation in digital CMOS circuits:

The most common is of CMOS scaling technology. The industry of semiconductor is slowly shifting towards more power consumption in the centric design technology. The total power dissipation in CMOS circuits (P_{Total}) is expressed the equation.

$$P_{total} = P_{switching} + P_{short-circuit} + P_{leakage} \qquad (2.1)$$

Where, P_{total} = Total power dissipation, Switching = Switching power, $P_{short-circuit}$ = Short circuit, Power $P_{leakage}$ = Leakage power

The total power dissipation is calculated with the help of this given as above equation.

2.3.2.1. Dynamic Power

The dynamic power dissipation in CMOS reversible logic gate is due to the charging and discharging of load capacitance (C_L) that is involved in the gate channel. This capacitance consists of the inside gate of logical circuit, the wire capacitance of the circuit which is dependent on the system fan out and the capacitance of the gate terminals that is using the transistor operational circuit, being the controlled fan-out of the net. The power dissipation can be calculated with the help of following equation [13], and as shown in fig.2.2.

$$P_{dyn} = C_L V_{dd}^2 f D \tag{2.2}$$

Where, P_{dyn}= Dynamic power dissipation of gate, C_L= Load capacitance of the gate, f= clock frequency, V_{dd}= supply voltage, D= Transition density of the output of the gate.

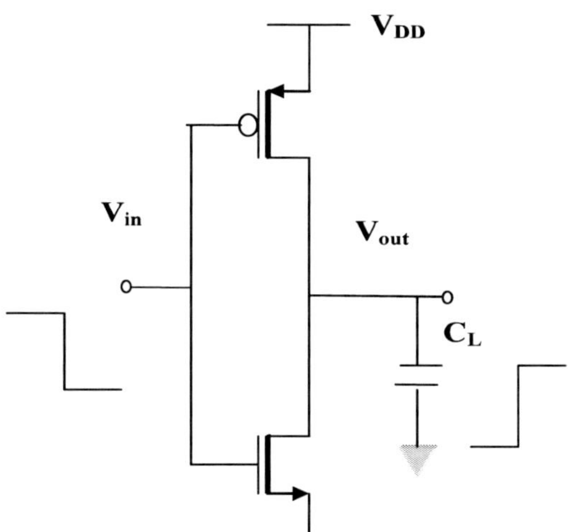

Figure 2.2. Dynamic Power Consumption.

The transition density is defined as the average number of transition occurring during one clock cycle. The dynamic power dissipation is proportional to the number of transitions occurring during the gate level. The accurate estimation of the transition density in reversible logic circuit will give an accurate estimation of dissipated power that power is the dynamic power. In the earlier technologies, the dynamic power which was mostly used in

CMOS logic design circuits [14]. But with the advancement of the Nanotechnology, other components of power consumption are also significant issues.

2.3.2.2. Short Circuit Power

The Short-Circuit power dissipation occurs during the switching activity into gate. During transition states this is short when both the NMOS and PMOS transistors are conducts. This is equivalent to shorting the supply V_{dd} and ground (Gnd) rails, for a brief period of time. The current flowing in this, through these transistors dissipates low power, as shown in the figure 2.3.

The value of the short circuit current is dependent on the operation of the capacitance load onto the output of the logical gate. In the high capacitance load, the output fall time is most of the time significantly larger than the input rise time. In this condition short circuit current is close to zero ($I_{sc}=0$) level of power. This conversely, for the low capacitive load, the fall time of the output is substantially smaller than the rise time transition. At this condition the short circuit current is maximum ($I_{sc}=I_{max}$). The short circuit power optimum techniques have the objective to make the output rise/fall time larger than the input rise/fall times. This amount of short circuit power is dependent on the switching speed of the logical gate operation. The amount of short circuit power dissipated into the inverter can be calculated by the formula:

$$P_{sc} = I_{sc}V_{dd} \tag{2.3}$$

Where, P_{sc} = Average Short-circuit power,
I_{sc} = Short-circuit current,
V_{dd} = Power supply

$$P_{short-circuit} = \frac{\beta}{12}\,(V_{dd} - 2V_T)^3\,\frac{\tau}{T} \tag{2.4}$$

Where, $P_{short\text{-}circuit}$ = short-circuit power dissipation,
β = Gain factor of the gate,
V_{dd} = supply voltage,
τ = Rise or fall time of input signal,
T = clock period

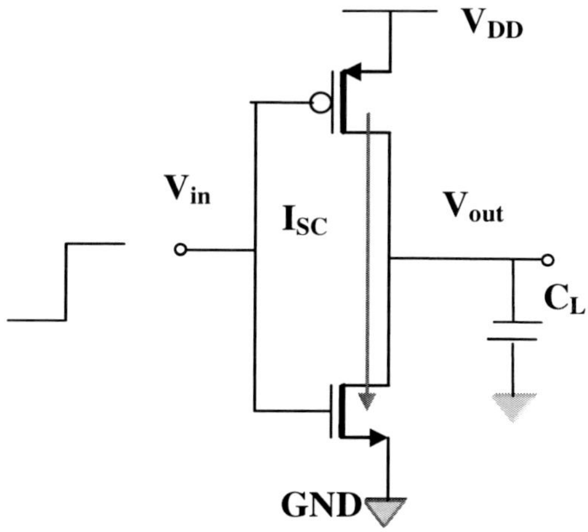

(a) Short-Circuit Power Dissipation.

The short-circuit power dissipation can be reduced to approximate 5 to 10% of the total power consumption of this reversible logic operational circuit by the proper sizing of the CMOS transistor.

CMOS transistor is determined to equal rise and fall times [15]. The three Figure 2.3 (a), (b), (c) shows the short-circuit power dissipation.

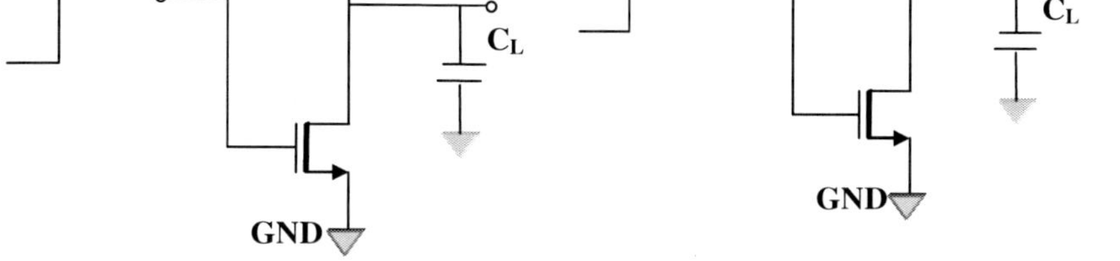

(b) Short-circuit power with large capacitive load, (c) short circuit power with small capacitive load.

Figure 2.3. (a) short-circuit power dissipation; (b) short-circuit power with large capacitive load, (c) short circuit power with small capacitive load.

2.3.2.3. Leakage (Static) Power Consumption

The quadratic dependence of the circuit is the switching power that is on the supply voltage that can utilize to reduce the switching power of the logic circuit [16]. In fact, when supply voltage is reduced, then this condition is also bound to the scaling of the device's sizing or the feature size. For example, when we reduce the gate oxide thickness limits, then the value of the maximum electric field that is tolerated on the gate terminal is imposing the upper bound on the supply voltage of device which it is using. Reducing the supply voltage forces the reduction of the threshold voltage of the device scaling in order to maintain the overdriving voltage (V_{GS} - V_{th}) of the circuit. That is a lead to the sub threshold of the leakage current that increases exponentially [17].

Indeed, the leakage power consumption is expected to dominate the overall power consumption as the scaling technology for low power issues. This is verified with the trend reflected in figure 2.4. where this leakage power will consume lower power, approximately 50% of the total power consumption for the technologies beyond 45 nm, if it is not properly addressed. The well known power dissipation is the one of the best requirement today in semiconductor industry. The power dissipation in the recent design of VLSI circuit is the low power design of the large level of reversible circuit. The leakage power consumption in the CMOS VLSI technology due to the sub-threshold current can be calculated using the expression.

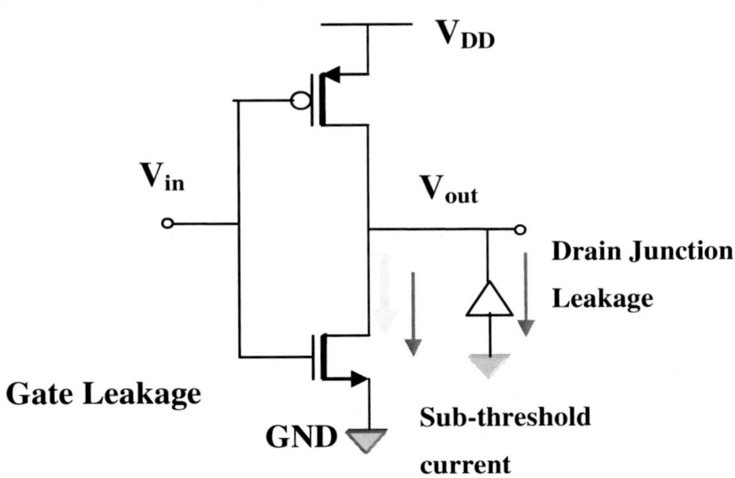

Figure 2.4. Leakage (Static) Power Consumption.

$$P_{leakage} = V_{dd} I_{sub} \tag{2.5}$$

Where, I_{sub} is the sub-threshold leakage current that is following the computation in the given expression.

$$I_{sub} = I_0 e^{\frac{(V_{gs}-V_{th})}{mV_T}} \left(1 - e^{-\frac{V_{ds}}{V_T}}\right) \tag{2.6}$$

Where, m = sub threshold swing coefficient, V_{th} is the threshold voltage, and $V_T = KT/q$ is the thermal voltage.

2.4. Basic GDI Cell (Gate Diffusion Input) Technique

This is a new low power design technique that allows solving the most critical problem with the help of this GDI cell technique. Transmission gate CMOS (TG) uses the transmission gate to realize the complex logic function using the small number of complementary metal oxide semiconductor. This problem solves the low level swing of using the MOS NMOS as well as PMOS [18].

The complementary metal oxide semiconductor pass-transistor logic (CPL) contains the features of the complementary input/output using the NMOS pass transistor logic with the reversible CMOS output inverters. Most of the features of CPL are the small stack height and the internal nod low power swing in the logical circuit, that contribute to the lowering the power consumption in the logic circuit. Pass transistor logic has suffered from the problem of static power consumption due to the low swing at the gate of the output inverters. The GDI cell approach allows implementation of a wide range of complex logic circuit function using only two transistors. This technique is most suitable for the design of fast low power circuits, used to reduce the number of transistors as compared to other CMOS existing pass transistor logic gate technique, while improving the logic level swing at the static power characteristics and allowing the simple top-down design by the using small cell design library. A simple change in the input configuration of GDI cell is based on Boolean function.

The basic concept of GDI cell technique is shown in figure 2.5. The PMOS in GDI cell is not connected to the supply VDD and in the other hand NMOS in GDI cell is not connected to GND. This feature gives the GDI cell

technique two extra input pins to use which makes the design more compatible in terms of no input increasing in the digital circuits. It must be remarked that not all the functions are possible in standard P-well process but the most successful design is in the CMOS process. One of the major problems of Gate-diffusion input technique requires a special CMOS process that the GDI schemes want twin-well CMOS or silicon on insulator (SOI) process to implement the design which makes it more expensive than the other standard cell [19]. One of the common basic problems is occurring in GDI cell based logic design of low swing of output signals because of the threshold voltage drop across the single channel pass transistor. A simple change can in the input configuration of GDI cell based on Boolean function that is different for every function. [20]. The reverse logic multiplexer with garbage input /output is more and the power dissipation in the circuits is less in comparison to the other CMOS logic reversible designs [21]. GDI Technique enables simpler reversible gates less transistor count and lower power dissipation in many implementations, as compared with other CMOS design techniques circuits in VLSI Design. The new device can have the basic wide spectrum of implementation in a wide variety of nanotechnology circuits. The future work of reversible logic multiplexer is modeling and simulation of the new Nano devices [22]. The basic GDI cell based functional table is given in 2.1. The function of F1 and F2 is the complete requirement of logic families. In this functionality, only F1 is the GDI function that can be realized for the standard P-well CMOS process, because the bulk of any NMOS is equally or constantly biased [23]. The GDI cell is the basic design technique for the high level circuit design of reversible logic gate.

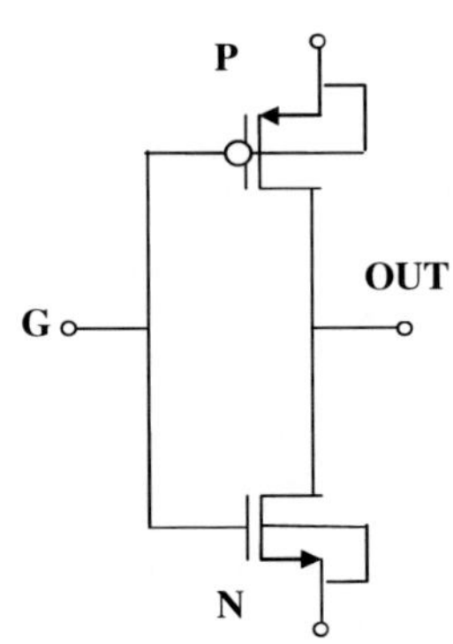

Figure 2.5. Basic Gate-Diffusion Input Cell.

Table 2.1. Functional table for Basic GDI Cell

N	P	G	OUT	Function
'0'	B	A	$\bar{A}B$	F1
B	'1'	A	$\bar{A}+B$	F2
'1'	B	A	A+B	OR
B	'0'	A	AB	AND
C	B	A	$\bar{A}B+AC$	MUX
'0'	'1'	A	$\bar{A}$	NOT

2.5. Leakage Reduction Techniques

As we move to the sub-micron technology, scaling of the transistor threshold voltage sharply increases the sub-threshold leakage current, whereas, the ultra thin gate oxide results in an exponential raise in gate leakage current. It is the current that flows during the protecting ground conductor to ground. In the deficiency of a grounding link, it is the current that could flow from any conductive part or the exterior of non-conductive parts to ground if a conductive path is presented (such as a human body). There are always unconnected currents flowing in the protection ground conductor. Many leakage reduction techniques have been proposed and their effectiveness and scaling trend have been evaluated [24]-[25]. The variations (such as transistor channel length and transistor threshold voltage) as a percentage of their nominal values increase when the technology advances. In [26], it is shown that the leakage current can vary from the target leakage current by 6.5 as when considering process variations. The measurement of chips in 0.18um technology exemplifies that 30mV variation in threshold voltage can result in 20 x differences in the leakage power and 30% variation in the frequency [27]. The techniques considered include increasing gate length [28], Vdd/Vth optimization [29], body biasing [30] and stack forcing [31].

2.5.1. Leakage Current Components

Sub Threshold Leakage Current: - It is the drain-source current of a transistor when the gate-source voltage is lower than the threshold voltage [32] and is mainly composed of diffusion current.

$$Isub = I0e^{\frac{vgs-vtho-\eta vds-\gamma vsb}{\eta v\theta}}\left(1-e^{\frac{-vds}{v\theta}}\right) \quad (2.6)$$

and

$$I0 = \mu cox\left(\frac{W}{L}\right)V\theta^2 e^{1.8}$$

W and L are the width and length of the transistor respectively while μ denotes carrier mobility.

$$V\theta = \frac{KT}{q}$$

V is the thermal voltage, θ is the Drain Induced Barrier Lowering (DIBL) coefficient and n is the slope shape factor/sub-threshold swing coefficient.

a) *Gate Leakage Current:* - Reduction of gate oxide thickness results in an augmentation in the electric field across the oxide. Thus, it leads to an exponential increase in tunneling probability of electrons through the gate oxide, and it means an exponential increase in the gate oxide tunneling current [33]. The gate leakage current is even larger than the sub-threshold leakage current from the 50nm process downwards [34].
b) *Junction Leakage Current:* - The reverse biased positive-channel, negative-channel (PN) junction leakage current has two main components: one corresponds to the minority carriers' diffusion near the edge of the depletion region, and the other takes place due to an electron-hole pair generation in the depletion region of the reverse biased junction. It is an exponential function of doping concentration and reverse biasing voltage across the junction.

2.5.2. Sources of Leakage Current

The diverse component of the leakage current is as follows.

- *PN junction reverse leakage:* this operation of the circuit occurs due to the minority carrier near to the depletion region. This current is a

small amount current and it can be considered as negligible current. However, if the electric field of the device reaches 10^6 V/m across the reverse biased PN junction, the current will flow due to Band-To-Band Tunneling (BTBT) [35].

$$I_{pn} = I_o \left(e^{(q/kT)V_{SB}} - 1\right) \quad (2.7)$$

Where, $I_0 = A.J_s$

A is the area of the junction and

J_s is the current density.

We need to keep the areas of the junction and channel facing sidewalls as small as possible, by minimizing the source and the drain areas.

- *Drain Induced barrier lowering:* during the relative size of drain/source, the depletion width of the vertical dimension size of the channel medium and the source drain potential of the device are affected to the band, bending over the significant width of the device channel. Higher potential of the drain is effected to reduce the threshold of the transistor, and increasing its leakage [36].
- *Gate induced drain Leakage:* The negative V_G and high V_D occur that causes the high electric field under the gate drain region overlapping ultimately causing the significant amount band-to-band tunneling.
- *Gate oxide tunneling:* This is due to tunneling in the electron through the gate oxide especially when under the condition when the electric field is very high coupled with the small oxide thickness.
- *Sub-threshold leakage:* that is the current flowing between the source and drain, when the condition stands that the transistor is in weak inversion and the gate voltage is below the threshold voltage. This is exponentially depending on the threshold voltage. Currently, it is the most difficult component to the leakage power consumption. The figure 2.6 is shown below.

2.6. Different Techniques to Reduce Leakage

Multi-Threshold and Power Gating CMOS

Multi-Threshold CMOS (MTCMOS) has emerged as a very popular technique for standby mode leakage power reduction [37]. In this technique, in

series with the power supply and the obtainable design and ground, a high-threshold voltage transistor is inserted. In fact, only one type of high *Vth* transistor is satisfactory for leakage reduction. The NMOS insertion idea is preferable, seeing that the NMOS on-resistance is smaller at the same width and therefore it can be sized smaller than an equivalent PMOS [38].

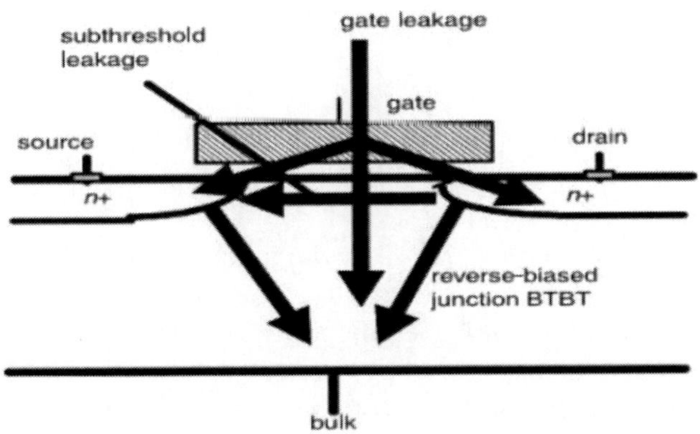

Figure.2.6. Leakage Current Component in a Transistor.

In MTCMOS technique, by using high threshold sleep transistor on the top and bottom of the logic circuit, transistors of low threshold voltage become detached from the power supply. Designing logic transistor having low threshold voltage (low-Vth) is illustrated in figure 2.7. Through the sleep signal, the sleep transistors are controlled. For the duration of the active, the sleep signal is disserted, making both high V_t transistor active and providing a virtual power and ground to inactivate Vt logic. When the circuit is inactive, sleep signal is asserted forcing both high V_t transistors inactive and detaching power lines from the low Vt logic. When the circuit is in standby mode, this results in a very low sub-threshold leakage current power to ground. One shortcoming of this method is that portioning and sizing of sleep transistors is complicated for large circuits [39]. The scaling process deviations become a vital concern in the earlier times. The technology is to reach nanometer establishment and it has resulted in rising leakage power dissipation [40].

Multi-threshold CMOS (known as power gating) is one of the frequently used technique to reduce standby leakage power. In MTCMOS technique, high-threshold voltage (High- V_{TH}) sleep transistor is used at the header and the footer of the circuit. Low-threshold (Low-V_{TH}) transistor is used to cut-off

the power supply. A PMOS transistor (High-V_{TH}) is placed between the actual power line and the virtual power line, a NMOS transistor (Low-V_{TH}) is placed between the actual ground and the virtual ground. The power consumption of a reversible logic has rapidly grown been the driving force to change to preferred choice of ICs [41].

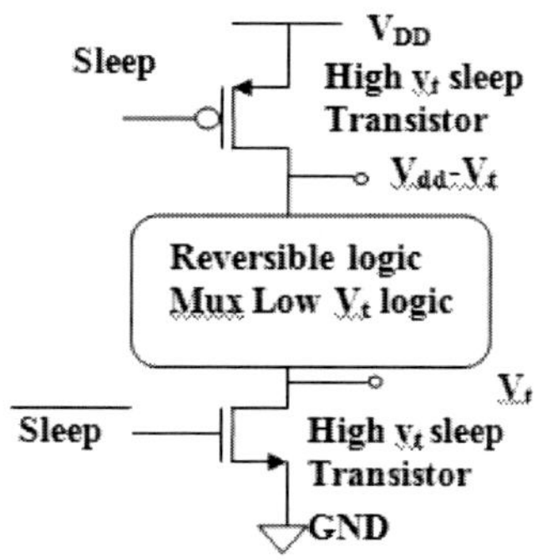

Figure 2.7. MTCMOS Technique Structure.

The leakage current flowing any logical circuit the stack of the connection of transistor is serially, reduces when more than the number of transistor of the stack is turned OFF. This effect is well known as the stacking effect [42]. When two or more than two transistors are OFF they are stacked on top of each other. Under this condition, low amount of leakage power is dissipated than when a single transistor is turned OFF.

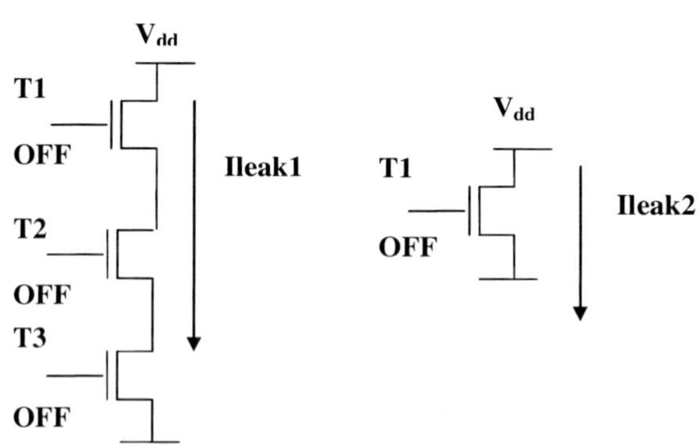

Figure 2.8. Transistor Stacking Effect.

This is the most proper region where each transistor in the stack induces a slight reverse bias between the gate and the source of the transistor below to the right. And that is increasing the threshold voltage of the bottom transistor making it more easily resistant to leakage. Therefore in the figure 2.8, transistor T2 leaks less amount of current than transistor T1 and T3 leaks less than the T2. In this condition the total amount of leakage current in transistors T1, T2 and T3 is diminishes as it flows from the supply voltage V_{dd} to the ground.

a) Super Cut-off CMOS Circuit

Another technique, which is an alternate to MTCMOS Power Gating, is Super Cut-off CMOS (SCCMOS) [43]. In this scheme, the sleep transistors are under-driven (or over-driven) when in standby mode. For example, an NMOS device would be turned OFF with a slight negative voltage instead of zero voltage. This means that the gate voltage is negative ($Vg < 0$).

b) Body Biasing

Reverse Body Biasing (RBB) has been used in memory chips since the 70's, to mitigate the risk of data destruction. In logic chips, on the other hand, the substrate and wells are typically biased stably to the ground and the power supply. However, since the 90's, reverse body biasing has been applied in logic chips for a different reason: power reduction. The original propose of the substrate biasing was utilized to reduce sub-threshold leakage during standby mode in portable applications.

In a circuit design, MOSFET is conventionally treated as a three-terminal device as shown in Figure 2.9. Where source and substrate terminal are shorted, it implies that the MOSFET operates as a zero-biased threshold device. On the other hand, when used as a four-terminal device, a forward or a reverse bias can be applied between the source and the substrate of a MOSFET. The MOSFET threshold voltage is improved, under the reverse body-bias. The MOSFET threshold voltage is decreased, under the forward body-bias. Figure 2.10 shows a standard CMOS inverter with shorted body-source transistors. Figure 2.9 displays, where the MOSFET is used as a four terminal device with a CMOS inverter under forward body-bias situation. The back gate forward body-bias method is shown in figure 2.9 is attuned with CMOS processes. Without any technology variation, the threshold voltage of MOS transistors can be concentrated electrically. A shortcoming of the bulk-driven technique is that it decreases the input impedance of an amplifier.

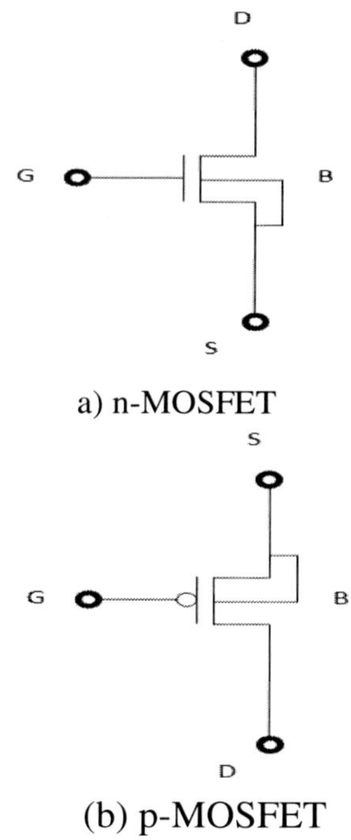

Figure 2.9. MOSFET as a three-terminal device (a) n-MOSFET (b) p-MOSFET. Here, S, G, D and B, respectively indicate source, gate, drain and substrate.

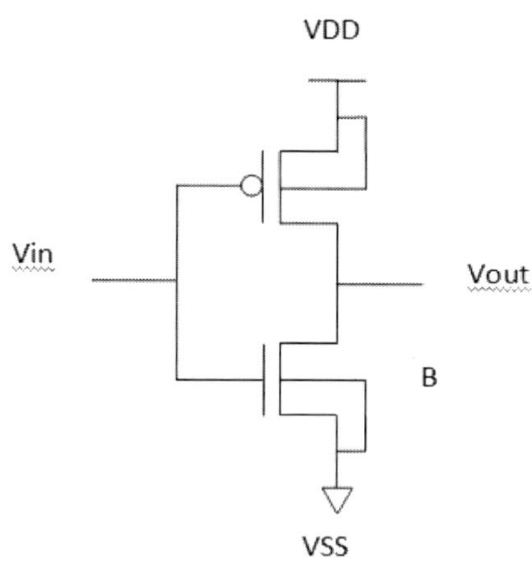

Figure 2.10. CMOS Inverter under Anti-Body-Source Short.

c) Variable Threshold CMOS (VTCMOS)

Variable threshold CMOS is a body biasing based implementation [44]. It uses a self-substrate bias circuit to control the body bias, to attain diverse threshold voltages. VTCMOS technique applies a Zero Body Bias (ZBB), in the active mode.

d) Dynamic Vth Scaling

To change Vth based on the performance demand, Dynamic Vth scaling (DVTS) method uses body biasing [45]. This technique uses the similar theory of earlier discussed VTCMOS technique. However, While DVTS modifies, its base on circuit performance demand, VTCMOS changes Vth based on both active and standby mode, i.e., the Vth is changed, based on the system recital condition.

e) Supply Voltage Scaling

Supply voltage scaling is used to reduce dynamic and leakage power. It was originally developed for switching power reduction. It is an effective method of consumption reduction due to the quadratic dependence of the switching power in relation to the supply voltage. Supply voltage scaling also provides leakage power savings. Lowering supply voltage provides an exponential reduction in sub-threshold current resulting from Drain-Induced Barrier Lowering (DIBL) effect. The DIBL effect tends to become more severe with process scaling to shorter gate lengths. For this reason, the achievable savings by this technique will increase with technology scaling.

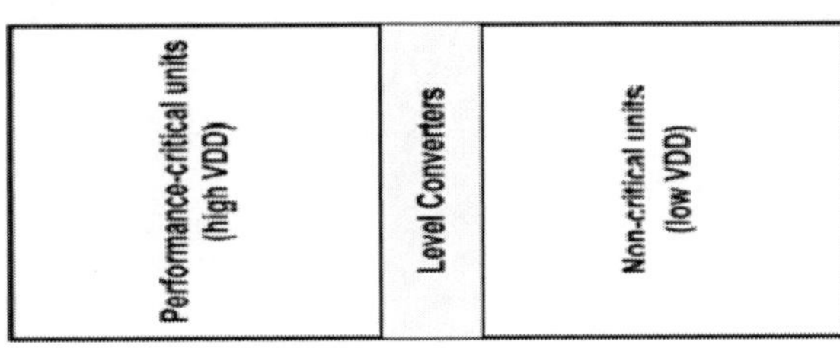

Figure 2.11. Two Level Multiple Static Supply Voltage Schemes.

f) Static Supply Scaling

In static supply scaling, multiple supply voltages are used as shown in figure 2.11. Critical and non-critical paths and/or units of the design are clustered and powered by higher and lower voltages, respectively [46]. In an extreme case, the combinational logic in a circuit can fall to zero when the

circuit is in idle mode because it does not need to hold its logic state, increasing the power saving.

g) Dynamic Supply Scaling

Dynamic supply scaling overrides the cost of using multiple supply voltages by adapting the single supply voltage to the performance demand. When performance demand is low, supply voltage and clock frequency are lowered, delivering reduced performance with substantial power reduction [47]. Dual threshold CMOS is a static technique that exploits the delay stack in non-critical paths to reduce leakage power. It provides both high and low threshold voltage (Vth) transistors in a single chip that are used to deal with the leakage problem.

h) Sleepy Stack

Sleepy stack is a technique to reduce the delay incurred by the forced staking of transistors. This is achieved by adding an additional transistor parallel to the additional stack transistor. This reduces the resistance offered during active operation. It's turned OFF when the circuit is idle.

2.7. Power and Delay Tradeoff in CMOS Circuit

After the design of logic signal the delay is found to be necessary in every circuit. Though this design, we need to find out the values of resistor that can introduce the delay on the signal lines. Delay of the circuit depends on the value of resistor itself as well as the capacitive load of the circuit. The propagation delay of the RC circuit can be calculated by the Elmore model delay formula as follows [48].

$$t_p = 0.69R_{eq} \times C_L \tag{2.8}$$

In the above equation, R_{eq} is the Amount of resistance that is implemented using the feed through cells. C_L is the Load capacitance that is seen by the reversible logic level gate. In most of the designs, the primary tradeoff is between power and delay. The delay of CMOS gate is time taken to charge and discharge the load capacitance $C_{L.}$ If we equalize the input, arrival times will reduce glitches and equalizing the edge rates will minimize the short circuit current. Clearly, it is not sufficient to optimize one without considering

the other. If we improve a design relative to power but it slows down the circuit, the result may not be acceptable. Then, the average gate delay for rising and falling transition is:

$$t_{avg} = \frac{t_{dr} + t_{df}}{2} \quad (2.9)$$

Where, t_{avg} = Average gate delay, t_{dr} = Rise time delay and t_{df} = fall time

Power delay Product is self-sufficient of switching frequency and can be calculated as

$$PDP = P \times t_d = 0.69V_{DD} \times \Delta V \times C \quad (2.10)$$

The delay in CMOS circuit gate is a function of V_{DD} and we can approximate it as:

$$\tau(V_{DD}) \propto \frac{C_L V_{DD}}{\beta (V_{DD} - V_T)^{\alpha}} \quad (2.11)$$

Where C_L is the load capacitance of the circuit, β is the unity gain factor and α is the velocity saturation Index. That models the short channel effect and takes the values.

The corresponding delay, when Power gating is known well, be τ (V_{DD}-ΔV). It is convenient to normalize the delay by τ (V_{DD}). We can also call it the Power Gating Delay (PGD) which is given below:

$$PGD = \frac{\tau(V_{DD} - \Delta V)}{\tau(V_{DD})} \approx \left(\frac{V_{DD} - V_T}{V_{DD} - \Delta V - V_T}\right)^{\alpha - 1} \quad (2.12)$$

It is provided that $V_{DD} \approx V_{DD} - V_T$. The voltage drop ΔV across the lower level in the circuit is the given values approximate.

$$\Delta V = RI = \frac{R_m}{W} \quad (2.13)$$

where R is the turn- off resistance of the lower transistor side which is equal to the turn- on resistance of a PMOS transistor of unit width divided by the width of the lower transistor width which is determined as the. Function of logical gate easy in terms of power gating.

$$W = \frac{1}{1-1-\alpha_{\sqrt{PGD}}}\left(\frac{R_m}{V_{DD}-V_T}\right)^I \quad (2.14)$$

2.8. Double Gate FinFET

On account of the poor electrostatic integrity of planar FETs and their rising susceptibility to process variations, technology scaling has exacerbated the twin problems of extreme leakage power and circuit consistency. Double-gate FETs are expected to substitute planar FETs, since the former have exceptionally short-channel uniqueness and their un-doped bodies are resistant to random dopant fluctuations. As the most suitable candidate amongst double-gate devices, FinFETs have emerged, sticking to their simplicity of fabrication. Dissimilar planar MOSFETs, the FinFET device structure consists of a vertical fin wrapped by electrically shorted or Independent Gates (IGs) on either side of the fin. In the Shorted-Gate (SG) mode of operation, the two gates are biased together to turn the device ON, akin to planar MOSFETs. However, in the IG mode, a bias on the back gate can alter the threshold voltage (Vth) of the front gate, thereby providing an effective method of controlling leakage power consumption at the expense of increased delay. Increasing logic density and reducing total power consumption by switching to FinFETs have gained substantial attention [49]. However, owing to densely packed features and [50], lithographic/process variations, FinFET circuits are very likely to suffer from a variety of physical defects [51].

As the Gate length in MOS devices is scaled down, suppression of the short-channel effects becomes more difficult. To overcome this problem, the incorporation of fin type MOSFETs (FinFETs) is an attractive option for the 22-nm node and beyond. Furthermore, the introduction of a Metal-Gate (MG) electrode is required for the FinFETs, not only to eliminate the gate-depletion effect but also to adjust threshold voltage Vth. In particular, titanium nitride (TiN) has been widely used for MG FinFETs because its midgap Work Function (WF) offers a symmetrical Vth for FinFET CMOS devices without channel doping. In principal, FinFETs with tall fins (large Hfin) can provide higher drive currents and smaller Vth variations than those with short fins (small Hfin) because of their increased channel width [52]. However, FinFET performance is affected by several factors, such as parasitic resistance Rp [53], channel stress due to the MG [54], and WF Variation (WFV) [55]. Thus, it remains unclear whether tall FinFETs are truly effective at boosting performance because these other factors are also influenced by Hfin [56]. In

this paper, we systematically investigated the influence of Hfin on MG-FinFET performance by fabricating n+-poly- Si/PVD-TiN stacked-gate FinFETs with different Hfin values [57]. Moreover, we will discuss the critical issues for realizing high performance FinFETs with increasing Hfin [58]. Figure 2.12. Shows schematic 3-D diagrams of the target FinFETs with different Hfin values. (001)-Oriented silicon on insulator (SOI) wafers were used as a starting material. We elaborate on the simulation setup used to simulate FinFET logic gates. We used FinFET as seen in figure 2.13, an environment that integrates double-gate compact models.

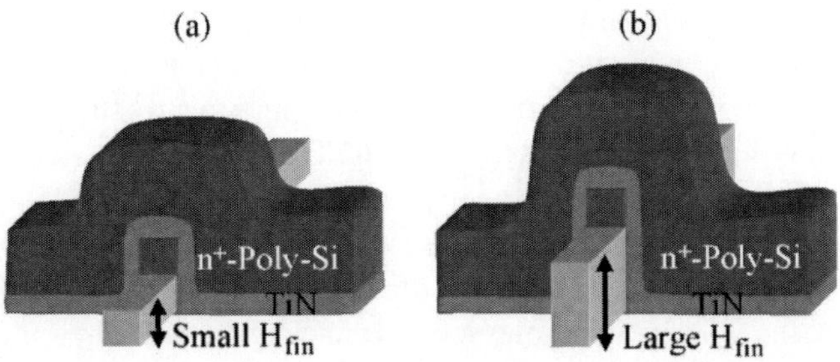

Figure 2.12. Schematics of the Fabricated FINFETs taking Hfin as a Parameter.

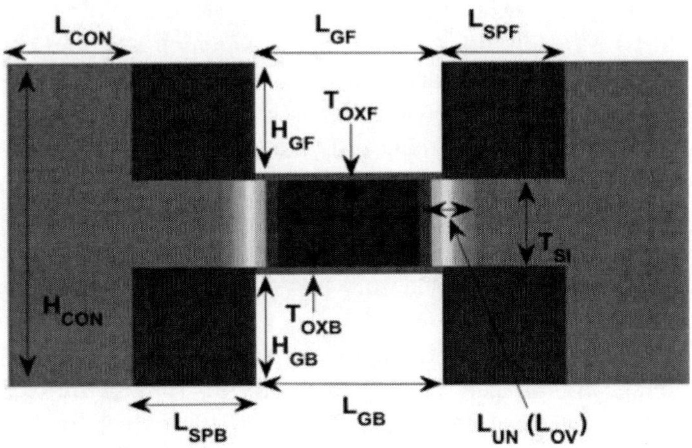

Figure 2.13. Simulated FINFET cross section.

REFERENCES

[1] Frank Poppen, *"Design Flow MicroPP Plus Documentation"*, (unpublished internal paper), 2000.

[2] G. L. Madhumati, Dr. M. Madhavilatha and K. Ramakoteswara Rao, Power and delay analysis of a 2-to-1 multiplexer implemented in multiple logic styles for multiplexer-based decoder in Flash ADC, *International Journal of Recent Trends inEngineering,* Vol 1, N0:4, 2009.

[3] G. E. Moore, *"Cramming More Components onto Integrated Circuits,"* *Electronics Magazine,* 38:114-117, 1965.

[4] *International Technology Roadmap for Semiconductors by Semiconductor Industry Association,* 2002. [Online]. Available http://public.itrs.net.

[5] Kim, N., Austin, T., Baauw, D., Mudge, T., Flautner, K., Hu, J., Irwin, M., Kandemir, M., And Narayanan, V., "Leakage Current: Moore's Law Meets Static Power," *IEEE Computer,* vol. 36, pp. 68–75, December 2003.

[6] J. Rabaey and M. pedrani. *"Low Power Design Methodologies."* Klinver- Acnderiiic Pirb/i.d7er~.B oston. MA. I996.

[7] M. Pedrani. "Power minimization in IC design: principles and applications," ACM Trnris. Dc.rigri Airtorifrrtio~i of Electronic Sv s t e~i i .v~o. l . I . Jan. 1996.

[8] M. Favalli and L. Benini. *"Analysis of glitch power dissipation in CMOS IC's."* Procrrrlir7g.r of I17ten~ntir~r~nl Svr~il~osiirr0i1i1 Loii3P oiivr Dcsigri. pp. 123-1 28. Apr. 1995.

[9] S. Rajagopal and G. Metha. *"Experiences with siniulationbased Schematic-level Power Estimation."* Procwdirigy of h7ffl77Afi<Jlln/ \'bJ-kShO/? OII LO\!' PO\i'eI' De.yig17. Pp.9- 14. Apr. 1994.

[10] Z. Liu, *"Multi-Voltage Nanoscale CMOS Circuit Techniques,"* Ph. D. Thesis, University of Wisconsin, Madison, pp.22-38, April 2008.

[11] R. Kumar, *"Temperature Adaptive and Variation Tolerant CMOS Circuits,"* Ph. D. thesis, University of Wisconsin, Madison, pp.1-8, May 2008.

[12] S. A. Tawfik *"Digital Circuit Methodology for low power and Robust Nanoscale,"* Ph. D. Thesis, University of Wisconsin, Madison, pp.3-25, 2009.

[13] G. E. Moore, "No exponential is forever: but "Forever" can be delayed!" *Proceedings of the IEEE International Conference Solid-State Circuits,* pp. 20-23, February 2003.

[14] H. Jiao and V. Kursun, "Reactivation noise suppression with sleep signal slew rate Modulation in MTCMOS circuits," *IEEE Transactions on Very Large Scale Integration (VLSI) Systems,* vol.21, no.3, pp.533-545, March 2013.

[15] M.S. Elrabaa, I.S. Abu-Khater, and M.I Elmasry, *"Advanced Low-power digital circuit techniques,"* Boston: Kluwer Academic publishers, 6th pp. 125-139, 1997.

[16] J.M. Rabaey and M. Pedram, editors, *"Low power design methodologies,"* Boston: Kluwer Academic Publishers, pp.367, 1996.

[17] H. Veendrik, "Short-Circuit dissipation of static CMOS circuitry and its impact on the design of buffer circuits," *IEEE Journal of solid state circuits,* vol. 19, no.4, pp.468-473, August 1984.

[18] L. Wei, K. Roy, and K. D. Vivek, "Low Voltage Low Power CMOS Design Techniques for Deep Submicron ICs," *Proceedings of the IEEE 13th International Conference on VLSI Design,* pp. 24–29, 2000.

[19] J. Kao, S. Narendra, and A. Chandrakasan, "Subthreshold Leakage Modeling and Reduction Techniques," *Proceedings of the IEEE/ACM International Conference on Computer-Aided Design (ICCAD'02),* pp. 141–148, 2002.

[20] N. Weste and K. Eshraghian, *"Principles of CMOS digital design,"* Reading, MA: Addison-Wesley 2nd edition, pp. 304–307, 2002.

[21] Ravish Aradhya H V, Praveen Kumar B V, Muralidhara K N "Design of Control unit for Low Power ALU Using Reversible Logic," *International Journal of Scientific & Engineering Research,* vol. 2, no. 9, pp.1-7, September-2011.

[22] Md. Saiful Islam, M. M. Rahman, Zerina Begum and Mohd. Zulfiquar Hafiz, "Low cost quantum realization of reversible multiplier circuit," *Information Technology Journal*, vol. 8, no. 2, pp. 208-213, Feb 2009.

[23] Rashmi S.B, Praveen B, and Tilak B.G., "Design of Optimized Reversible BCD Adder/Subtractor," *International Journal of Engineering and Technology (IJET),* vol. 3, no. 3, pp. 230-234, May 2011.

[24] Prathibha Rani T.R, H.R. Bhagyalakshm, M.K. Venkatesha, "Optimized Reversible carry select BCD adders using Reversible logic gates," *International Journal Computer Technology & Application (IJCTA),* vol.3, no.4, pp.1550-1556, July-August 2012.

[25] Mohammad Samadi Gharajeh and Majid Haghparast "On Design of a Fault Tolerant Reversible 4-Bit Binary Counter with Parallel Load," *Australian Journal of Basic and Applied Sciences,* vol.6, no.7, pp. 430-446, 2012.

[26] Y-F. Tsai, et al., "Implications of Technology Scaling on Leakage Reduction Techniques", *Design Automation Conference,* pp.187-190, Jun. 2003.

[27] B. Chatterjee, et al., "Effectiveness and Scaling Trends of Leakage Control Techniques for Sub-130nm CMOS Technologies", *International Symp. On Low Power Electronics and Designs,* pp. 122-127, Aug 2003

[28] S. Nerendra, et al., "Full-Chip Subthreshold Leakage Power Prediction and Reduction Techniques for Sub-0.18-um CMOS", *IEEE Journal of Solid-State circuits,* VOL. 39, No. 2, pp. 501-510, Feb. 2004.

[29] Bokar, et al., " Parameter Variations and Impact on Circuits and Microarchitecture", *Design Automation Conference,* pp.338-342, June 2003 and leakage Under the Presence of Process Variation", IEEE Transaction of VLSI Systems, Vol. 11, No. 5, pp.888-899, Oct. 2003

[30] N. Sirisantana, et al., "High-Performance CMOS Circuits using Multiple Channel Length and Multiple Oxide Thickness", *International Conference on Computer Design,* pp. 227-232, Sep. 2000

[31] L. Wei, et al., "Design and Optimization of Low Voltage High Performance Dual Threshold CMOS Circuits", *Design Automation Conference,* pp. 489-494, Jun. 1998.

[32] T. Chen, et al., *"Comparison of Adaptive Body Bias (ABB) and Adaptive Supply Voltage (ASV) for Improving Delay".*

[33] Ye Y., et al., "A New Technique for Standby Leakage Reduction in High-Performance Circuits", *Symp. On VLSI Circuits,* pp. 40-41, Dec 1998.

[34] B. Amelifard, F. Fallah and M. Pedram "Leakage minimization of SRAM cells in a dual -Vt and dual –tox, technology", *IEEE transactions on Very Large Scale Integration (VLSI) systems,* vol.16, pp. 851-9, Jul 2008.

[35] K. Roy, S. Mukhopadhyay, and H. Mahamoodi-meimand,"Leakage Current mechanisms and Leakage reduction techniques in deep sub-micrometer CMOS circuits", *Proceedings of the IEEE,* vol.91, pp. 305-327, Feb 2003.

[36] A. Agarwal, S. Mukhopadhyay, A.Raychowdhury.K. Roy, and C.H.Kim, "Leakage Power Analysis and reduction for nanoscale circuits", *Micro IEEE,* vol 26, pp. 68-80,2006.

[37] Y. Taur and T. H. Ning, "*Fundamentals of Modern VLSI Devices,*" New York: Cambridge University Press, New York, pp.194-200, 1998.

[38] K. Roy, S. Mukhopadhyay, and H. Mahmoodi-Meimand, "Leakage Current Mechanisms and Leakage Reduction Techniques in Deep-Submicrometer CMOS Circuits," *Proceedings of the IEEE,* vol. 91, no. 2, pp. 305–327, Feb. 2003.

[39] Mutoh, S., Douscki, T., Matsuya, Y., Aoki, T., Shigematsu, S., and Yamada, J. 1-V Power Supply High-Speed Digital Circuit Technology with Multithreshold-Voltage CMOS. *IEEE Journal of Solid-State Circuits* 30, 8 (Au- gust 1995), 847-854.

[40] KAO, J. et al. Transistor Sizing Issues and Tool for Multi-Threshold CMOS Technology. In: *Design automation conference,* 1997. Proceedings... [S.l.]: IEEE, 1997, p.409-414.

[41] Hao Xu, Ranga Vemuri, and Wen-Ben Jone, Senior Member, IEEE "Dynamic Characteristics of Power Gating During Mode Transition," *IEEE Transactions On Very Large Scale Integration (VLSI) Systems,* vol. 19, no. 2, pp.237-249, February 2011.

[42] Kanak Agarwal, Harmander Deogun, Dennis Sylvester, Kevin Nowka, "Power Gating with Multiple Sleep Modes," *This work was supported in part under Defence Advanced Research Project Agency Contract,* University of Michigan, pp.1-5,2012.

[43] Adil Zaidi, Kapil Garg, Ankit Verma, Ashish Raheja "Design & Simulation of CMOS Inverter at Nanoscale beyond 22nm," *International Journal of Emerging Science and Engineering (IJESE),* vol. 1, no.5, pp.83-87, March 2013.

[44] Linfeng Li and Jianping Hu, "A Transmission Gate Flip-Flop Based on Dual-Threshold CMOS Techniques," *Proceedings of IEEE International Symposium on Circuits and Systems (ISCAS),* pp. 539-542, 2009.

[45] Kawaguchi, H., Nose, K., and Sakurai, T. A Super Cut-Off CMOS (SCCMOS) Scheme for 0.5-V Supply Voltage with Picoampere Stand-By Current. *IEEE Journal of Solid State Circuits* 35, 10 (October 2000)

[46] KURODA, T. et al. A 0.9-V, 150-MHz, 10-mW, 4 mm2, 2-D Discrete Cosine Transform Core Processor with Variable Threshold-Voltage (VT) Scheme. *IEEE Journal of Solid State Circuits,* New York, v.31, n.11, p. 1770-1779, Nov 1996.

[47] Kim, C. H.; ROY, K. Dynamic Vth Scaling Scheme for Active Leakage Power Reduction. In: *Design automation and test in europe conference,* 2002. Proceedings IEEE, 2002, p. 163-167.

[48] Takahasi, M. et al. A 60-mW MPEG4 video codec using clustered voltage scaling with variable supply-voltage scheme. *IEEE Journal of Solid State Circuits,* New York, v.33, n.11, p. 1772-1780, Nov. 1998.

[49] Burd, T. D. et al. A Dynamic Voltage Scaled Microprocessor System. *IEEE Journal of Solid State Circuits,* New York, v.35, n.11, p. 1571-1580, Nov. 2000. Dual Threshold CMOS.

[50] J.M. Rabaey, A. Chandrakasan, and B. Nikolic, "Digital Integrated Circuits: A design Perspective", 2nd edition, *Prentice Hall Electronics and VLSI series Upper saddle River,* NJ: Pearson Education, pp.411-413, 2003.

[51] A. Muttreja, N. Agarwal, and N. K. Jha, "CMOS logic design with independent-gate FinFETs," in *Proc. Int. Conf. Comput. Design*, Oct. 2007, pp. 560–567.

[52] P. Mishra, A. Muttreja, and N. K. Jha, "Low-power FinFET circuit synthesis using multiple supply and threshold voltages," *J. Emerging Tech. Comput. Syst.*, vol. 5, no. 2, pp. 1–23, 2009.

[53] A. Datta, A. Goel, R. Cakici, H. Mahmoodi, D. Lekshmanan, and K. Roy, "Modeling and circuit synthesis for independently controlled double gate FinFET devices," *IEEE Trans. Comput.-Aided Design Integr. Circuits Syst.*, vol. 26, no. 11, pp. 1957–1966, Nov. 2007.

[54] K. Itoh, "Adaptive circuits for the 0.5-V nanoscale CMOS era," in *Proc. ISSCC Tech. Dig.*, 2009, pp. 14–20.

[55] T. Matsukawa, Y. X. Liu, K. Endo, J. Tsukada, Y. Ishikawa, H. Yamauchi, S. O'uchi, K. Sakamoto, and M. Masahara, "Influence of NiSi on parasitic resistance fluctuation of FinFETs," in *Proc. Symp. VLSI-TSA*, 2011, pp. 1–2.

[56] T. Matsukawa, K. Endo, Y. Ishikawa, H. Yamauchi, S. O'uchi, Y. X. Liu, J. Tsukada, K. Ishii, K. Sakamoto, E. Suzuki, and M. Masher, "Fluctuation analysis of parasitic resistance in FinFETs with scaled fin thickness," *IEEE Electron Device Let.*, vol. 30, no. 4, pp. 407–409, Apr. 2009.

[57] T. Kamei, Y. X. Liu, K. Endo, S. Ouchy, J. Tsoumada, H. Yamauchi, Y. Ishikawa, T. Hayshed, T. Matsukawa, K. Sakamoto, A. Ogura, and M. Masher, "Experimental study of physical-vapour-deposited titanium nitride gate with an n+-polycrystalline silicon capping layer and its application to 20 nm fin-type double-gate metal–oxide–semiconductor

field-effect transistors," *Jon. J. Appl. Phys.*, vol. 50, no. 4, pp. 04DC14-1– 04DC14-5, Apr. 2011.

[58] T. Matsukawa, S. Ouchy, K. Endo, Y. Ishikawa, H. Yamauchi, Y. X. Liu, J. Tsoumada, K. Sakamoto, and M. Masher, "Comprehensive analysis of variability sources of FinFET characteristics," in *VLSI Sump. Tech. Dig.*, 2009, pp. 118–119.

Chapter 3

Various Logic Style of Multiplexer for Low Power Designs

3.1. Objectives of the Chapter

For low-power, Very Large Scale Intigration (VLSI) can be assigned to special design levels, such as the behavioural, structural, layout and the process technology level are the escalating requirements. Multiplexer is the basic component consisting of CMOS (Complementary Metal Oxide Semiconductor) memory elements and data control structures. The low-power consumption is one of the most significant concerns of the system design.High speed applications developed different styles for low power designs.

The purpose of this chapter is to reduce power and area of 2:1 MUX, 4:1 MUX and Reversible logic while maintaining the viable performance. The diverse configurations are designed using different topology of 2:1 MUX and 4:1 MUX such as CMOS based MUX, transmission gate and pass transistor. Multiple logic styles are used to design a low power 2:1 multiplaxer and 4:1 multiplexer (MUX). The leakage current, power consumption, delay and transistor count up has been compared for different logic styles of 2:1 MUX and 4:1 MUX. Reversible logic has emerged as one of the most important approaches and a more prominent technology having its applications in low power CMOS, garbage inputs/outputs, communication, nanotechnology and computer graphics. This chapter proposes a novel reversible multiplexer gate and the design of differential reversible multiplexer using the proposed reversible gate is also discussed.

3.2. MULTIPLEXER

In the design of large-scale digital systems, it quite frequently happens, that a single line is required to carry two or more diverse digital signals. Obviously, only one signal at a time can be placed on one line. It is required front a device that will allow us to select, at dissimilar instants, the signal we desire to place on this common line. Such a circuit is referred to as Multiplexer. The graphical symbol and truth table of 4:1 MUX is presented in figure 3.1, (a) and (b) respectively. A multiplexer performs the function of selecting the input on any one of 'n' input lines and feeding this input to one output line. Multiplexers are used as one technique of declining the number of integrated circuit packages requisite by a particular circuit design. Then, it reduces the cost of the system.

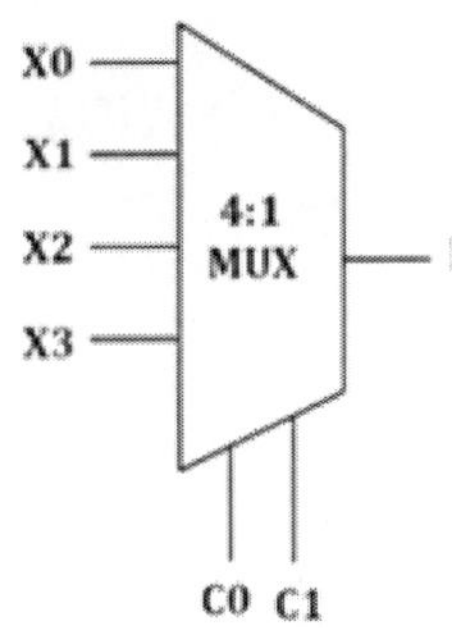

(a)

C0	C1	M
0	0	X0
0	1	X1
1	0	X2
1	1	X3

(b)

Figure 3.1. Graphical Symbol 4.1 MUX.

$$\text{Output}=X0.\overline{C_0}.\overline{C_1}+X1.\overline{C_0}.C1+X2.C0.\overline{C_1}+X3.C0.C1$$

Assume that we have four lines, X0, X1, X2 and X3, which are to be multiplexed on a single line, Output (M). The four input lines are also identified as the Data Inputs. While there are four inputs, we will require two extra inputs to multiplexer, recognized as the Select Inputs, to select which of the X inputs is to appear at the output, called as select lines C0 and C1. The gate implementation of a 4:1 MUX is depicted in figure 3.2.

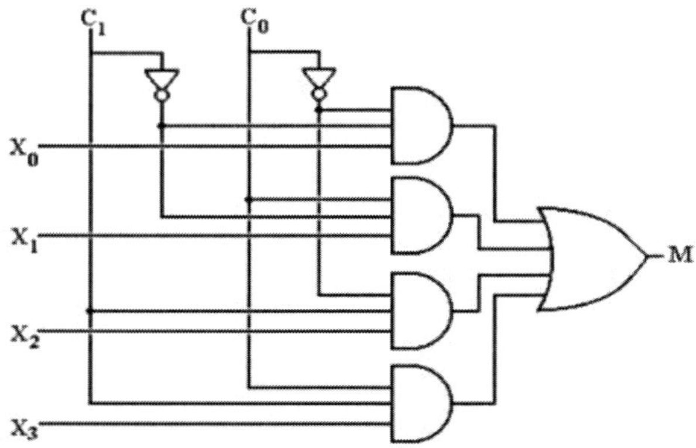

Figure 3.2. Gate Implementation of a 4.1 Multiplexer.

One of the number of data inputs, that a circuit generates an output exactly reflects the state, depends on particular value of one or more control inputs called "multiplexer". A multiplexer with two data inputs is called as '2 to 1 or 2:1' multiplexer. Usually used circuit and logical symbol for 2:1 multiplexer is illustrated in figure.3.3. Logic expression for multiplexer output is given below.

$$F = \hat{S}\,x1 + S\,x2 \qquad (3.1)$$

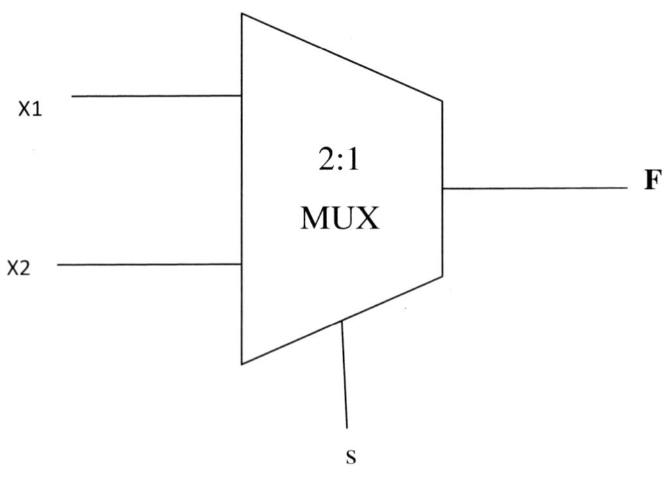

Figure 3.3. Graphical Symbol 2:1 MUX.

3.3. Realization of Various Logic Styles of Multiplexer

For wireless sensor network applications, the supply voltage may be reduced below the threshold voltage due to lower frequency rate. This operation is referred to as sub-threshold design that uses the sub-threshold current as constrain current to estimate the inputs [1]. Logic Style is the representation of the logic function from a set of transistors. It impacts the speed, size, power dissipation and wiring difficulty of a circuit. This uniqueness may differ significantly from one logic style to another and thus make the proper choice of logic style crucial for circuit performance. Different 2:1 and 4:1 MUX topologies have been proposed for reliable Sub-threshold applications.

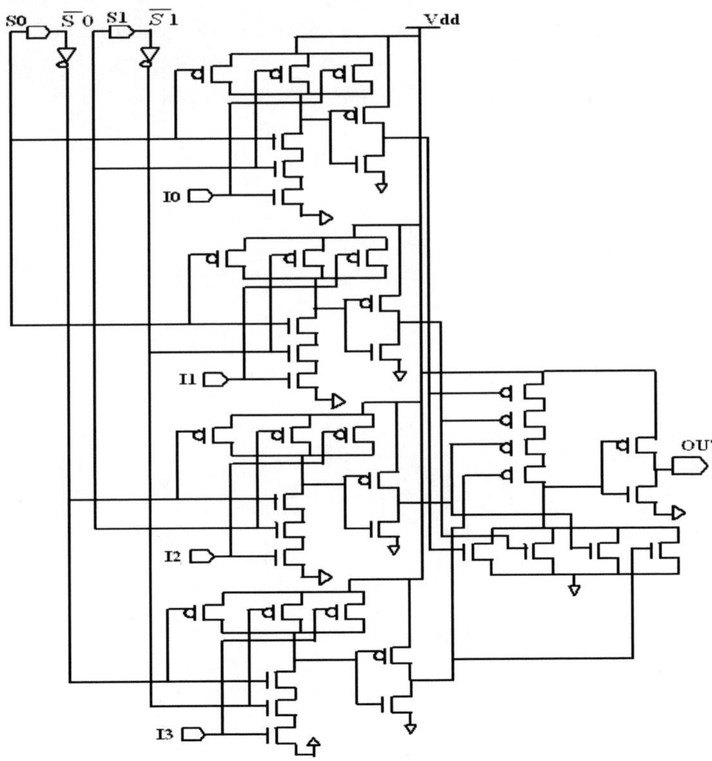

Figure 3.4. Static CMOS 4.1 MUX.

There are different static logic styles like CMOS, Pseudo, PTL, DPL, LEAP, TR Gates which exist. Depending on the requirement of parameters like leakage power, delay and area, the particular logic styles can be selected.

3.3.1. Static CMOS MUX

The static 4:1 MUX transistor level representation consists of 44 transistors. Due to excessive transistors, the required functionality will use more power with abundant area. But as the technology is falling, the circuit should perform with minimum power consumption. As a result, this design is not measured more advantageous for lower technology. This circuit is transistor level architecture of 4:1 MUX, which consists of 42 transistors.

The power consumption is reduced with minimum area in NMOS and transmission gate MUX design. The Static CMOS MUX representation of a 4:1 multiplexer is illustrated in figure 3.4.

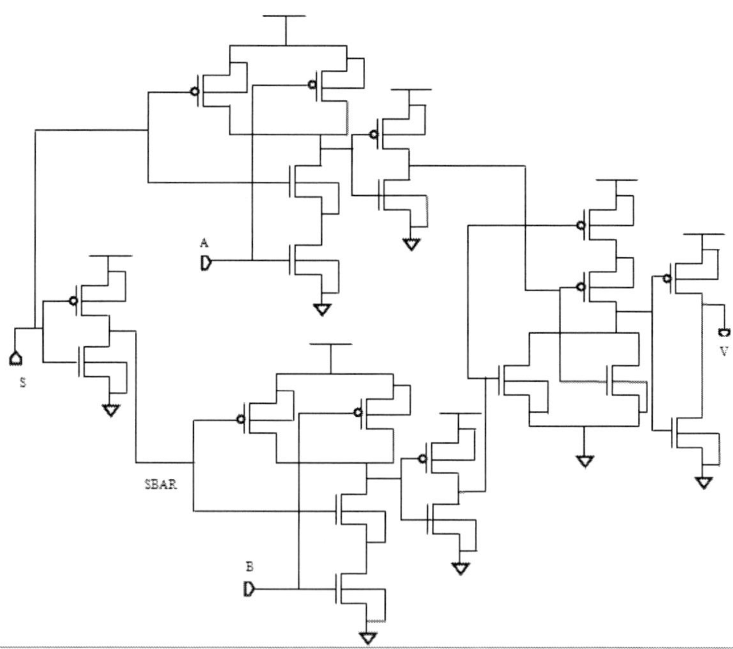

Figure 3.5. Schematic of CMOS based 2:1 MUX.

A simple realization of 2-to-1 MUX would require 2 AND gate, an OR gate and a NOT gate. Schematic of CMOS based 2:1 MUX is given in figure 3.5. Any logic function in complementary CMOS is implemented by NMOS pull-down and PMOS pull-up networks associated between gate, output and power lines [2].

3.3.2. Transmission Gate Logic MUX (TGL-Gate)

This design is the transmission gate type of MUX structure realized with minimum number of transistors compared to conventional CMOS based design. The transmission gate combines the best of both the properties by placing NMOS in parallel with the PMOS device. Two transmission gates perform 2:1 MUX function. So, three pairs of 2:1 MUX are required to form a transmission gate 4:1 MUX structure. The transmission gate based 4:1 MUX demonstrated in figure 3.6. An advantage in this MUX design is the remarkable gain in terms of transistor's count as compared to static MUX with minimum power consumption but compromises on delay.

This is the transmission gate based 2:1 MUX structure implemented with very minimum transistors (4 MOS transistors) compare to the CMOS based 2:1 MUX which has 20 CMOS devices. The back to back connected PMOS & NMOS transistors arrangement acts as a switch termed ‘Transmission Gate’. In transmission gate, NMOS transistor passes a strong 0, but a weak 1, while PMOS transistor passes a strong 1, but a weak 0. The CMOS based 2:1 MUX uses NMOS transistor act as pull down network and PMOS transistor act as pull up network, whereas the transmission gate combines both the properties by placing NMOS transistor parallel with the PMOS transistor. Two transmission gates are connected as shows the schematic of transmission gate based 2:1 MUX in figure 3.7 to form a MUX structure [3]. Each the transmission gate acts as an AND switch to replace the AND logic gate which is used in a CMOS based design of MUX. Hence, the transistor count is reduced to 4 and it shows that it occupies less area as compared to CMOS based MUX. One more change when compared to CMOS based 2:1 MUX is that there is no supply voltage applied to the circuit. It results in less operating power. It has lower gate delay and the circuit propagates faster than that of the CMOS based 2:1 MUX. The gate delay as mentioned can be calculated thus as:

$$t_{pd} \propto C_L V_{dd} / I_{DS} \tag{3.2}$$

where, t_{pd} = propagation delay,
CL = load Capacitance,
V_{dd} = supply voltage and
I_{DS} = drain saturation current

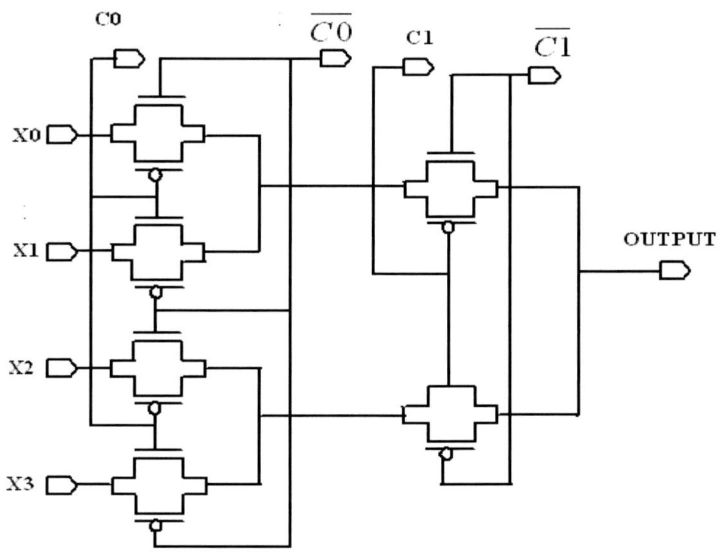

Figure 3.6. Transmission Gate based 4.1 MUX.

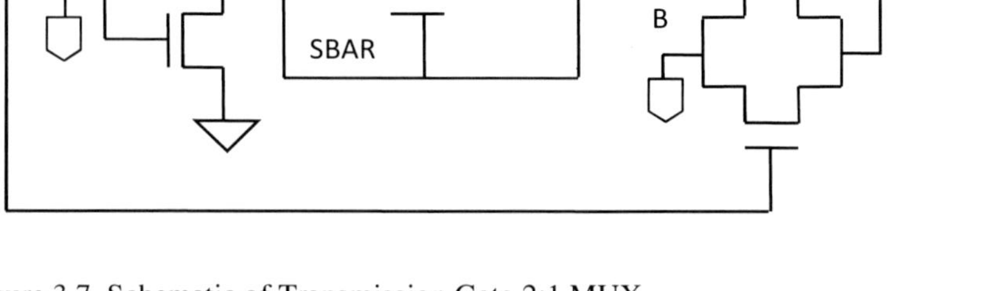

Figure 3.7. Schematic of Transmission Gate 2:1 MUX.

3.3.3. Pass-Transistor (PTL) MUX

Static CMOS is slowed because an input must drive both NMOS and PMOS transistors. The number of transistors gets reduced in pass-transistor logic, required to implement logic to drive gate terminals as well as source drain terminals by allowing the primary inputs. To perform the logic operation, one pass-transistor network (either NMOS or PMOS) is sufficient and the greatest advantage of pass transistor logic [4]-[5].

Among all these NMOS, Multiplexer is optimal. It uses 8 NMOS transistors and these eight-pass transistors in the input select which signal to propagate. The logic levels will be deteriorated by the pass transistor. The schematic figure of NMOS pass-transistor logic is presented in figure 3.8

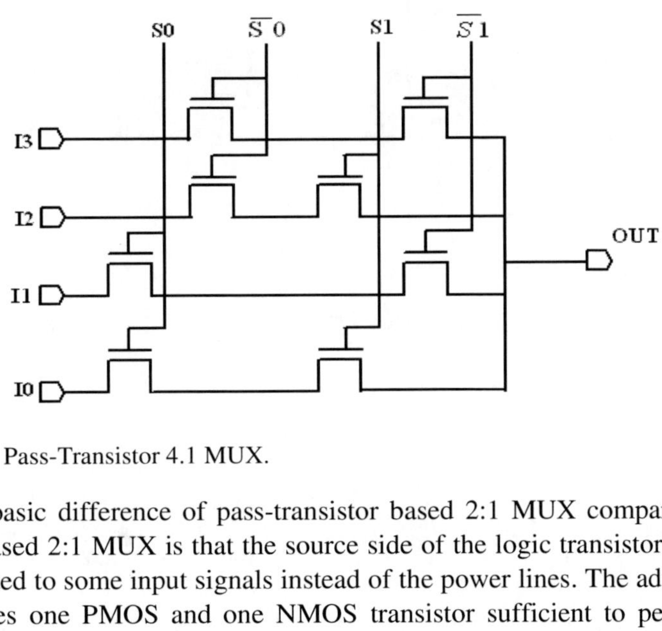

Figure 3.8. Pass-Transistor 4.1 MUX.

The basic difference of pass-transistor based 2:1 MUX compared to the CMOS based 2:1 MUX is that the source side of the logic transistor networks is connected to some input signals instead of the power lines. The advantage is that it uses one PMOS and one NMOS transistor sufficient to perform the logic operation, which results in a minimum number of transistors (2 MOS transistor) and these two pass transistors at the input select which signal to propagate, especially when NMOS networks are used. The Schematic of Pass Transistor based MUX is displayed in figure 3.8. The logic levels will be deteriorated by the pass transistor. The threshold voltage of both pass-transistors should be one and the same for accurate operation. However, the threshold voltage drops (Vout= Vdd-Vtn) through the NMOS transistors while

passing a logic '1' makes move backward and forward (or level) return at the gate outputs, necessary in order to avoid constant currents at the consequent output inverters or logic gates. Adjusting the threshold voltages (i.e., Vtp>Vtn) as a solution at the process technology level is usually not sufficient for other reasons. Among all these topologies of 2:1 Multiplexer, it is optimal.

It is more efficient and compared to other topologies occupies less area. Finally, pass-transistor logic with move backward and forward re-establishment circuitry is sensitive to voltage scaling and transistor sizing with respect to circuit robustness (reduced low power and area occupied), i.e., efficient or reliable operation of logic gates is not essentially certain at low voltages or small transistor sizes. Several pass-transistor logic styles are available like Complementary Pass Transistor Logic (CPL), Energy Economized Pass Transistor Logic (EEPL) [6].

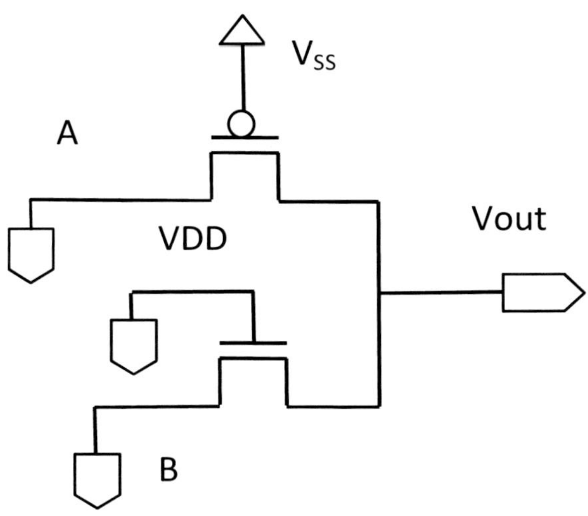

Figure 3.9. Schematic of Pass Transistor 2:1 MUX.

3.3.4. Pseudo NMOS MUX

Static CMOS are slowed because an input must drive both NMOS and PMOS transistors. Pseudo NMOS gate resembles static gates, but substitute the slow PMOS pull up stack with a particular grounded PMOS transistor which acts as a pull up resistor. The effective pull up resistance should be large enough so that the NMOS transistors can pull the output near to ground,

yet low enough to rapidly pull the output high. The schematic diagram of pseudo NMOS 4:1 MUX is shown below in figure 3.10.

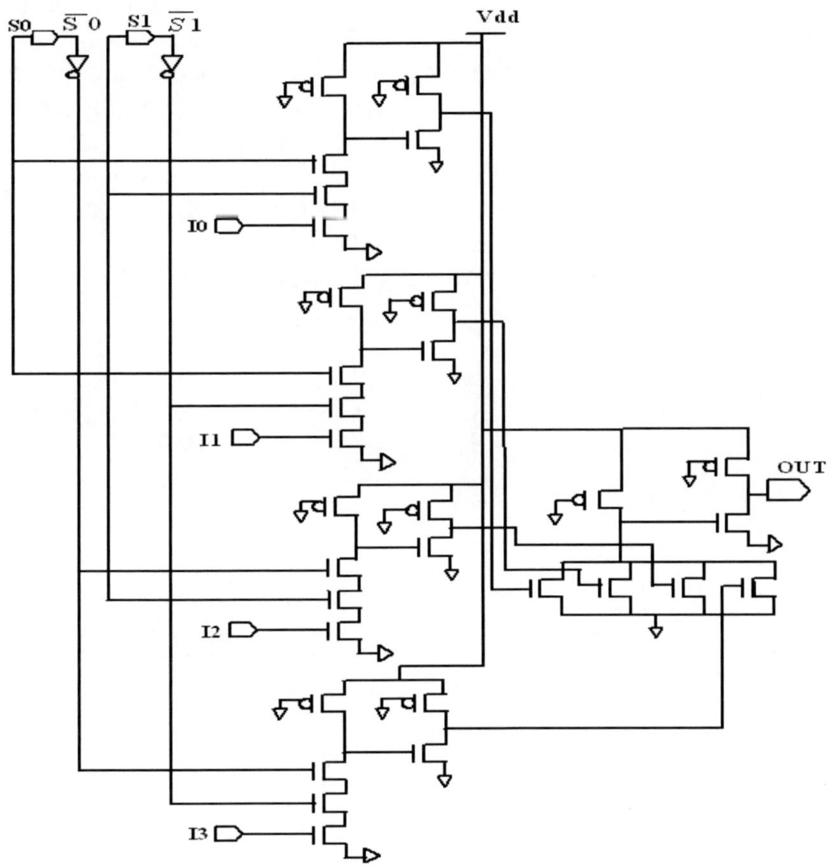

Figure 3.10. Pseudo NMOS Transistor 4.1 MUX.

3.3.5. Leap MUX

In NMOS PTL schematic, the feedback of PMOS is used which is called a level restorer. This level restorer is used to store degraded logic ‘1’ from output of NMOS pass transistor. The schematic diagram of LEAP 4:1 MUX is brought in below figure 3.11.

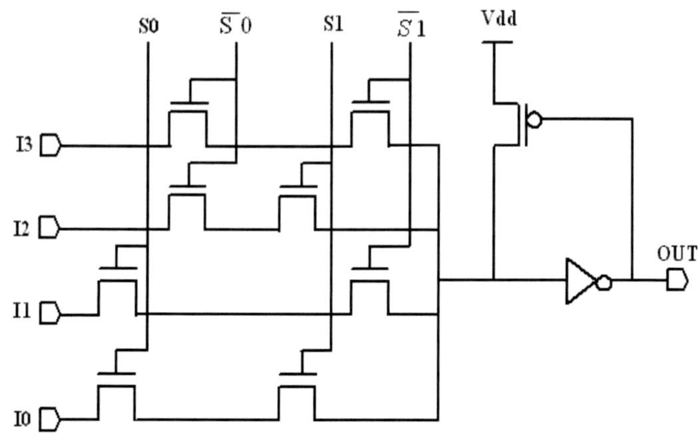

Figure 3.11. LEAP 4.1 MUX.

3.4. Reversible Logic

Reversible logic is a very prospective approach of logic synthesis for power reduction in future computing technologies [7]. In a reversible logic circuit, the number of inputs and outputs are same and there is a one-to-one mapping between input and output values. Reversible circuits are constructed using reversible gates. The recent computers erase a bit of information every time they perform a logical operation. Such logical operations are generally known as 'irreversible logic'. It imposes many design constraints. Zero energy dissipation would be possible only if the network consists of reversible gates, so that reversibility becomes an essential property in future circuit design [8]. According to Landauer [9], traditional irreversible hardware computation inevitably leads to energy dissipation due to the loss of each one bit of information which dissipates an amount of KTln(2) Joules of energy, where K is the Boltzmann's constant (1.3807×10^{-23} JK^{-1}) and T is the absolute temperature at which computation is performed. In the case of high technology circuits, the heat that is dissipated is very large, when compared to a loss of one bit information at room temperature. Furthermore, voltage-coded logic signals have an energy of $E_{sig}=1/2CV^2$ and this energy is dissipated whenever the node voltage changes in the irreversible CMOS technology. It is estimated that reversible logic circuits also help to secure energy by using charge recovery logic [10]. As the Moore's law continues to hold, the processing

power just doubles every 18 months. The present irreversible technologies will dissipate a lot of heat and can reduce the life of the circuit. The reversible logic operations do not lose (erase) information and dissipate very less heat. Thus, reversible logic circuits are likely to be in demand in high-speed power aware circuits. Reversible logic circuits are of high interest in low-power CMOS design, nanotechnology, quantum computing and optical computing. The most important application of reversible logic lies in quantum computers [11]. A quantum computer will be viewed as a quantum network (or a family of quantum networks) composed of quantum logic gates; each gate performing an elementary unitary operation on one, two or more than two–state quantum systems known as Qubits. Each Qubit representing an elementary unit of information; corresponding to the classical bit values '0' and '1'.

3.4.1. Basic Reversible Logic Gate

A reversible logic gate is n × k logic device where n and k are the number of inputs and outputs with one-to-one mapping, i.e. n should be equal to k (n = k). A gate is said to be reversible if the input vector can be uniquely recovered from the output vector and there is a one-to-one correspondence between its input and output logic [12]. A set of reversible logic gates is needed to design reversible logic circuits [13]. An n × k reversible logic gate can be represented as:

$$I_V = (I_1, I_2, I_3 \dots \dots \dots I_n) \tag{3.3}$$

$$O_V = (O_1, O_2, O_3 \dots \dots \dots O_k) \tag{3.4}$$

Where, I_v and O_v are input and output vectors. Consider the following issues to perform the synthesis of reversible gates and obtain optimization. Any reversible circuit realizes only the function that is reversible. The number of outputs in a reversible gate or circuit are same as the number of inputs, and generally used traditional NOT gate is the only reversible gate. Every reversible gate has a cost associated with it, called Quantum cost. The simplest reversible gate is NOT gate and is a 1×1 gate. The controlled NOT (CNOT) gate is an example of a 2×2 gate. Many of the available gates are 3×3 Reversible gates such as TR, FG, PG and TG gates. The Quantum Cost of 1×1 Reversible gates is zero, and the Quantum Cost of 2×2 Reversible gates is one. Any reversible gate is optimized by using 1×1 NOT gates (inverter) and 2×2

(Feynman gate) Reversible gates, such as V, V^+ (V is square root of NOT gate and V^+ is its Hermitian) and FG gate that is also known as CNOT gate. The properties of V and V^+ Quantum Gates are given in the following equation [14].

$$V \times V = NOT \tag{3.5}$$

$$V \times V^+ = V^+ \times V = I \tag{3.6}$$

$$V^+ \times V^+ = NOT \tag{3.7}$$

The Quantum Cost of a Reversible gate is calculated by counting the number of V, V^+ and CNOT gates. One of the most important features of a reverse gate is its garbage output, i.e. each and all input of the gate that is not used as input to another gate or as a primary output is called garbage output.

The constants are used as the input line that is either set to Zero (0) or One (1) in the circuit input side. Gate Count is the number of reversible gates used to realize the function. The Hardware Complexity refers to the number of basic gates (NOT, AND and EXOR gate) used with ease to synthesize the given more functions [15]. The delay can be calculated in reversible logic based on the critical path. It is the longest path that gives the desired output of the system [16].

There exist many reversible gates in literature. Among them Feynman gate, (FG) [17], Peres gate (PG) [18], Toffoli gate (TG) [19], Fredkin gate (FRG) [20] and Khan Gate (NG) [21] are most common. The quantum realizations of all these gates are not available in the literature. Only PG, FG, TG and FRG have been realized in nanotechnology. A reversible logic circuit should have the following features. These are the various condition for reversible logic circuit.

- Least number of reversible gates
- Least number of constant inputs
- Minimum number of garbage outputs
- Length of cascading gates should be least
- Same number of outputs as that of inputs

3.4.2. Design of Basic Reversible Logic Multiplexer Using Pass Transistor

We design reversible multiplexer circuits using dual-line pass transistor logic and monotone circuit [22]. Boolean values Z = 1 (one) and Z = 0 are represented by (Z, Z) = (1, 0) and (Z, Z) = (0, 1) respectively. For example, Reversible Inverter is shown in figure 3.12. It consists of a metal cross-over connection. Because of the monotone circuit, we set all values (Z, Z) = (0, 0).

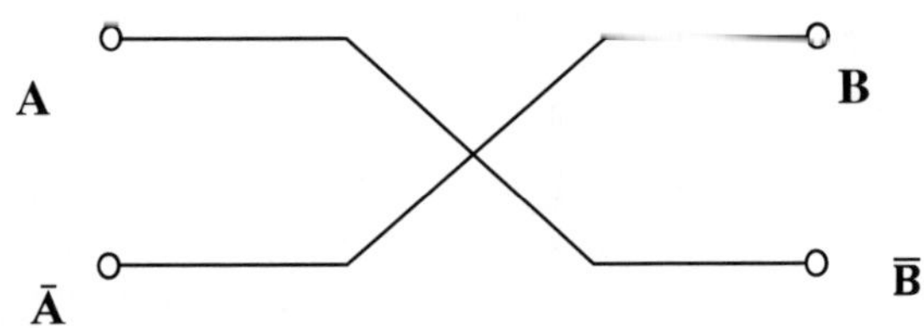

Figure 3.12. Reversible Inverter.

For the implementation of an 'on-off' switch in a circuit, we use two way CMOS transmission gate with two control inputs as figured in 3.13.

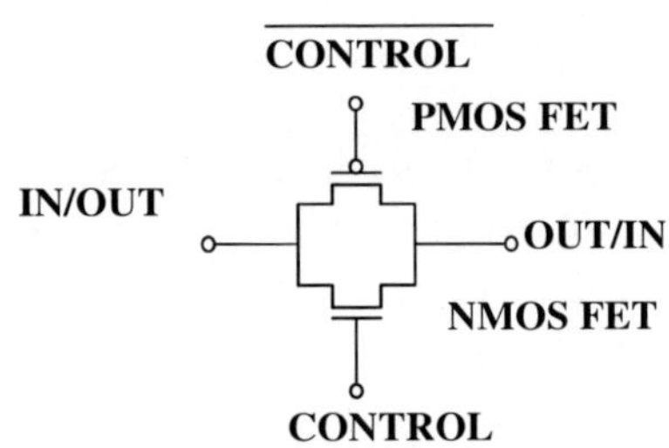

Figure 3.13. CMOS Transmission Gate.

3.5. Proposed Reversible Multiplexer

3.5.1. 2:1 Reversible Multiplexer using Pass Transistor

We design 2:1 Reversible Multiplexer using pass transistor which has two inputs A, B and one select input S, the output of the circuit is Y as shown in

figure 3.14. We can implement many combination circuits using this 2:1 Reversible multiplexer, adder, subtractor, multiplier, etc.

Pass transistors are used here to design the circuit voltage reduction in present circuit [23]. Functionality of Two transistors multiplexer shown below is the tabled 3.1.

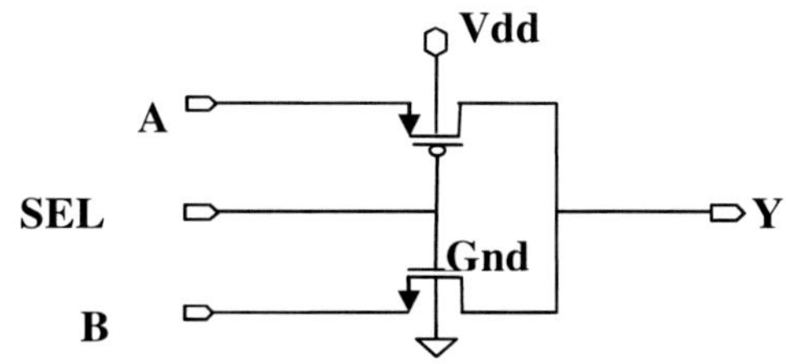

Figure 3.14. 2:1 Multiplexer uses Pass Transistor.

Table 3.1. Functional Table of 2:1 Reversible Multiplexer

Select	Output
0	A
1	B

It demonstrates their transient response of 2:1 multiplexer using cadence virtuoso tool using 45nm Technology where, power supplies Vdd 0.7 volts. The simulation results in terms of waveform of leakage power response is 0.0193 (nW), Average power 0.05044 (nW), delay 9.69 (ns) and Power Delay Product (PDP) 4.883×10^{-18} Joule.

3.5.2. 2:1 Basic Concept of Design 2:1 Reversible Multiplexer Function

The proposed reversible 2:1 multiplexer gate is realized using transistor implementation as described in figure 3.15. The required output Y can be obtained using only two transistors. However, three transistors are required for calculating garbage outputs. To obtain output G1, a pass transistor is used for passing the selection input S (select) to the output G1 as detailed figure 3.15 (c). This design shows the response output of 2:1 reversible logic gate multiplexer. This operation of reversible logic gate circuit required less number of transistor count.

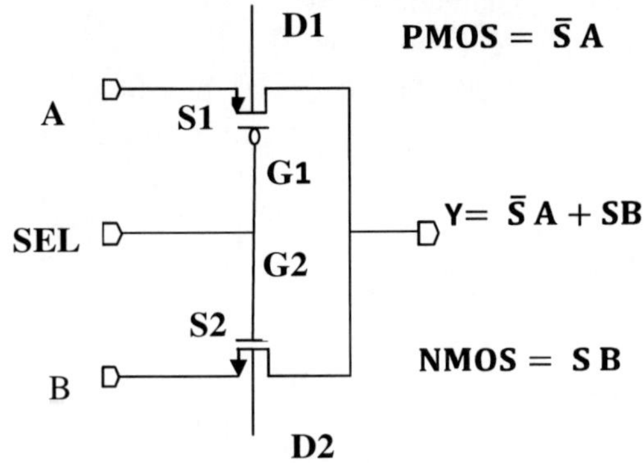

(a) Circuit Diagram for Multiplexed Output Y

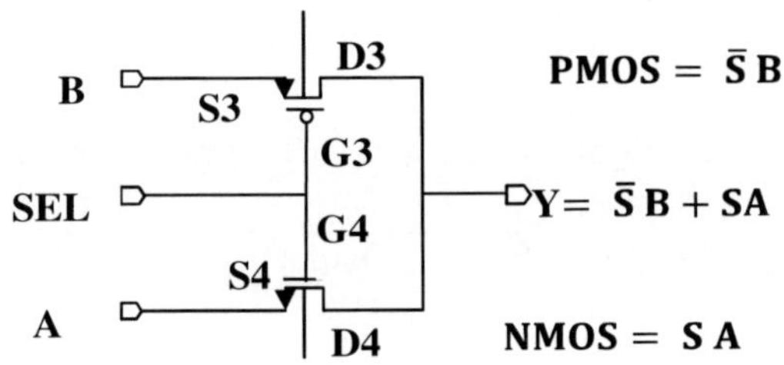

(b) Circuit Diagram for Output G2

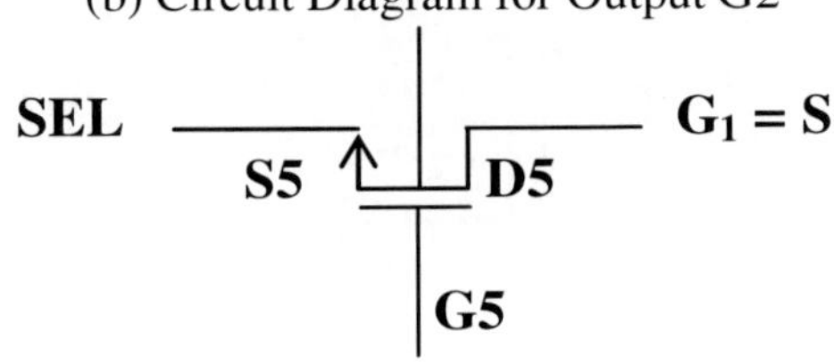

(c) Circuit Diagram for Output G1

Figure 3.15. Implementation of reversible multiplexer with garbage input/ output Gate (a) Circuit diagram for multiplexed output Y (b) Circuit diagram for output G2 (c) Circuit diagram for output G1.

In figure 3.15, when the selected input is ‘0’, the PMOS transistor will be ON while NMOS will be ‘off’ thus input A is passed to the output Y. If selection input is 1, then the NMOS transistor will be conducting while PMOS is in OFF state, passing input B to the output Y. Similar to output Y, output G2 is passed with input B when S=0 (SEL) and input A, when S=1[24].

3.5.3. 2:1 Reversible Multiplexer with Garbage Input/Output

This paper proposed reversible multiplexer gate is proposed in order to function as the 2:1 reversible multiplexer producing two garbage bits.

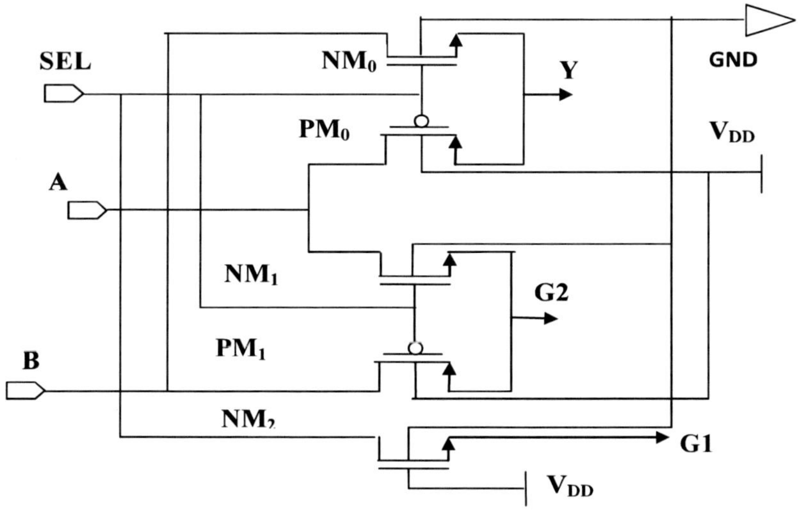

Figure 3.16. Transistor Implementation of 2:1 Reversible Multiplexer with Garbage Input/Output.

The input s are A, B, S and it depends on the selected input S, the corresponding information bit is passed on to the output Y in figure 3.16.

Table 3.2. Truth Table of 2:1 Reversible Multiplexer with Garbage Input/Output

Input			Output		
SEL	A	B	G1	Y	G2
0	0	0	0	0	0
0	0	1	0	0	1
0	1	0	0	1	0
0	1	1	0	1	1
1	0	0	1	0	0
1	0	1	1	1	0
1	1	0	1	0	1
1	1	1	1	1	1

The table 3.2 shows the various input values of the circuit i.e. A, B, Sel (S) and the output values G1, Y and G2. The output of this reversible logic circuit is shown here. The response of the output and all the levels of the output of this design depend on the select line. If select line is high so is the output.

The table 3.2 show the input and output values with select line and input A and B, the output Y with two garbage output G1 and G2 and the transient response of the reversal of the 2:1 Reversible multiplexer with garbage input/output. The simulation results in terms of waveform of leakage power response is 0.000000016 (nW), average power 1.421 (nW), delay 9.852 (ns) and Power Delay Product (PDP) 2.384×10^{-25}.

This figure3.17 depicts that the delay comparison in two technologies 180 nm and 45 nm generate the simulation result as given below. It is observed in this simulation delay decreases in estimate 180 nm technology in 45 nm, the Power Delay Product is also much better in this operation. The design is very useful for low power and small amount of delay in the circuit.

The voltage gain of reversible multiplexer with garbage input/output calculated after simulation is 760mv (0.76). After the comparative study of two technologies, results convey 45 nm to 180 nm in terms of leakage power, average power, delay as well as PDP.

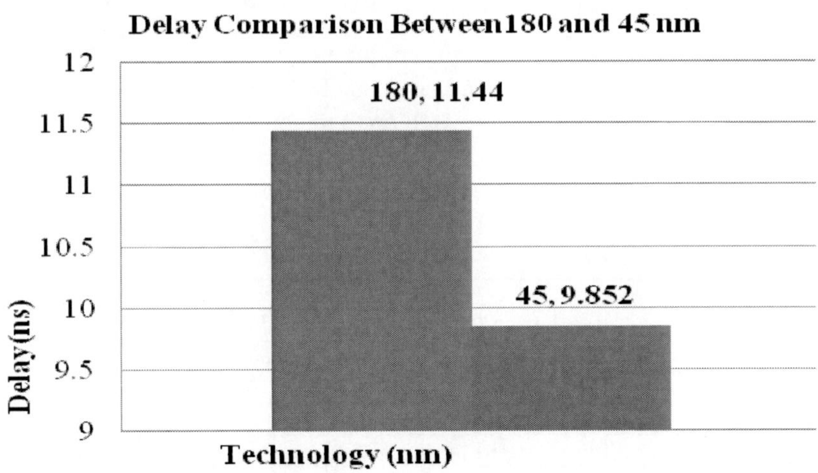

Figure 3.17. Delay Comparison 2:1 Reversible Multiplexer with Garbage Input/Output.

REFERENCES

[1] A. Wang and A. Chandrakasan, "A 180mV FFT processor using sub threshold circuit techniques", *IEEE,* ISSCC, pp.292-529, 2004.

[2] E. R. Hsieh and S. S. Chung, "The proximity of strain induced effect to improve the electron mobility in a silicon–carbon source–drain structrue of n-channel metal–oxide semiconductor field-effect transistors," *Appl. Phys. Lett.,* vol. 96, no. 9, p. 093501, Mar. 2010.

[3] K. Akarvardar, C. D. Young, D. Veksler, K.-W. Ang, I. Ok, M. Rodgers et al., "Performance and variability in multi -VT FinFETs using fin doping," in *Proc. VLSI TSA,* 2012, to be published.

[4] K. Yano, Y. Sasaki, K. Rikino, and K. Seki, Topdown pass-transistor logic design, *IEEE Journal of Solid-State Circuits, vol. 31,* June 1996, 792–803.

[5] Radhakrishnan D, Whitaker S.R, Maki G.K, Formal design procedures for pass transistor switching circuits, *IEEE Journal of solid state circuits,* vol-20. No 2, April 1985, 531-536.

[6] P. Zuber, M. Miranda, M. Bardon, S. Cosemans, P. Roussel, P. Dobrovolny, T. Chiarella, N. Horiguchi, A. Mercha, T. Y. Hoffmann, D. Verkest, and S. Biesemans, "Variability and technology aware SRAM Product yield maximization," in *VLSI Symp. Tech. Dig.,* Jun. 2011, pp. 222–223.

[7] T. Toffoli., *"Reversible Computing," Automata, Languages and Programming, Journal* W. Bakker and J. van Leeuwen, ED. Springer Verlag, vol. 85, pp. 632-644, 1980.

[8] H. Bennett, *"Logical reversibility of computation," IBM Journal Research and Development*, vol. 17, no.6, pp. 525-532, 1973.

[9] Landauer, R., 1961. "Irreversibility and heat generation in the computing process," *IBM Journal Research and Development,* vol. 5 no.3, pp. 183-191, July 1961.

[10] M. P. Frank. "Introduction to reversible computing: motivation, progress, and challenges," *Proceedings of the 2nd Conference on Computing Frontiers, Italy and New York,* USA, pp. 385–390, May 2005.

[11] Vlatko Vedral, Adriano Bareno and Artur Ekert, *"Quantum Networks for Elementary Arithmetic Operations,"* vol. 1, pp. 102-109, Nov 1995.

[12] Perkowski. M, Jozwiak. L, Kerntopf. P, Mishchenko. A, Al-Rabadi.A, Coppola.A, Buller.A, Song.X, Khann.M.M.H.A., Yanushkevich.S,

Shmerko.V, and Chrzanowskas-Jeske.M.2001 "A general decomposition for reversible logic," *Proceeding of 5th International Reed-Muller Workshop,* Starwille, pp. 119-138, 2001.

[13] Hafiz Md. Hasan Babu and Chowdhury. A.R, "Design of Reversible Binary coded Decimal adder by using Reversible 4- bit parallel Adder," *Proceeding of IEEE 18th International Conference on VLSI Design,* pp. 255-260, 2005.

[14] W.N. N. Hung, X. Song, G. Yang, J. Yang, and M. Perkowski, "Optimal Synthesis of Multiple Output Boolean Functions Using a Set of Quantum Gates by Symbolic Reach ability Analysis," *IEEE Transactions on Computer-Aided Design of Integrated Circuits and System,* vol. 25, no.9, pp.1652-1663, September 2006.

[15] Md. Saiful Islam, M. M. Rahman, Zerina Begum and Mohd. Zulfiquar Hafiz, "Low cost quantum realization of reversible multiplier circuit," *Information Technology Journal,* vol. 8, no. 2, pp. 208-213, Feb 2009.

[16] Rashmi S.B, Praveen B, and Tilak B.G., "Design of Optimized Reversible BCD Adder/Subtractor," *International Journal of Engineering and Technology (IJET),* vol. 3, no. 3, pp. 230-234, May 2011.

[17] R. Feynman, "Quantum mechanical computers," *Optic News,* vol. 11, pp.11-20, 1985.

[18] A. Peres, "Reversible logic and quantum computers," *Physics Reversible,* pp. 3266-3276, 1985.

[19] T. Toffoli, "Reversible computing," Tech Memo MIT/LCS/TM-151, *MIT Lab for Computer Science.,* vol.85, pp.632-664, 1980.

[20] Fredkin and T. Toffoli, "Conservative logic," *International Journal of Theoretical Physics,* vol.21, no.12, pp.219-253, 1982.

[21] M.H.A. Khan, "Design of full-adder with reversible gates," *Proceeding of 5th International Conference on Computer and Information Technology,* Dhaka, Bangladesh, pp. 515-519, 2002.

[22] Alex De Vos, *"Reversible computing," Progress in Quantum Electronics,* vol. 23, no.1, pp. 1-49, 1999.

[23] Praveen B, Rashmi S.B, Tilak B.G, "Transistor Implementation of Reversible PRT gates," *International Journal of Engineering Science and Technology (IJCER),* vol. 3, no. 3, pp. 2289-2297, 2011.

[24] Praveen.B, Vinay Kumar S.B "2^n:1 Reversible Multiplexer and its Transistor Implementation," *International Journal of Engineering Science and Technology (IJCER),* vol. 2, no.1, pp.182-189, Jan-Feb 2012.

Chapter 4

LEAKAGE REDUCTION TECHNIQUES ON MULTIPLEXER CIRCUIT FOR THE SYSTEM SOC DESIGN

4.1. OBJECTIVES OF THE CHAPTER

Technology scales and leakage currents grow exponentially and become an increasingly large component of total power dissipation. In this chapter, influence of leakage reduction techniques on multiplexer/load circuit is analyzed. The low-power consumption is one of the most important issues for the system SOC design. Different techniques and technologies for low-power designs in high speed interface applications are developed and also applied in the practical design projects. In order to suppress the degradation of signals and to enhance the operation speed, we designed interconnection for the circuit using Self- Controllable Voltage Level (SVL) techniques. The designed circuit is realized in a High variability standard 45nm process and uses 0.7V supply voltage. Our optimization technique, using the proposed method, reduces the power consumption and leakage the current by significant amount of multiplexer /load circuit. Body bias technique which decreases the process variation on MUX is applied on MUX to decrease low power, leakage current and bulk voltage. We investigate the use of body biasing to adapt to control the bulk voltages of transistor and show that it is particularly effective in sub-threshold circuits and can eliminate performance variations with low power.

4.2. Mechanisms of Leakage Reduction Techniques

In this section, two major circuit design techniques namely, MTCMOS and VTCMOS for leakage reduction in digital circuits are described:

4.2.1. MTCMOS (Multi-Threshold CMOS)

In the recent research of the reduction of sub-threshold leakage is targeted, because the gate leakage of the circuit is still in small orders of magnitude than the sub- threshold leakage for the current generation of the Nanoscale technologies. Since the sub-threshold leakage current is exponentially dependent on the threshold voltage of gate oxide, when increasing the threshold voltage, would decrease the leakage current substantially. Most threshold voltage devices have larger threshold switching delays.

The various leakage current mechanisms of this work and several leakage reduction techniques in the CMOS logic circuits were discussed in [1]. The several leakage related techniques for reducing leakage power involve the static and dynamic approaches for the design of the reversible logic multiplexer [2]. The dynamic approach involves various decisions of the runtime dependent making for the leakage reduction into the reversible logic. The most popular issue today is the leakage reduction technique using the sleep transistor in the MTCMOS circuits for controlling the leakage power of the circuit [3]. Various optimizations for the MTCOMS circuits involving the basic use of the sleep transistor have been proposed such as [4]. Most of the technique can reduce the standby leakage power and the performance degradation of logic level design of low power CMOS reversible logic [5].

The basic MTCMOS structure is shown in figure 4.1, where a low Vth computation block is gated with high Vth power switches. When the high Vth transistors are turned ON, the low Vth logic gates are connected to the virtual ground and the power, and switching is performed through fast devices [6]. A specialized case of dual Vth technology that is more effective at reducing leakage currents in the standby mode is MTCMOS (Multi-Threshold CMOS). Though, by introducing an extra series device to the power supplies, MTCMOS circuits will provoke a performance penalty compared to CMOS circuits, which declines if the devices are not sized large enough. When the circuit enters the sleep mode, the high Vth gating transistors are turned OFF, resulting in a very low leakage current from V_{CC} to the ground. MTCMOS is only efficient at reducing standby leakage currents and therefore, is most

effective in burst mode type application, where reducing standby power is a major benefit.

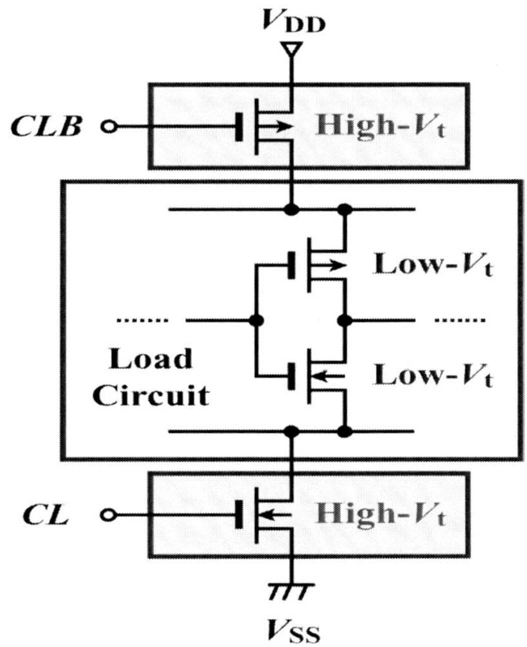

Figure 4.1. MTCMOS.

4.2.2. VTCMOS (Variable Threshold CMOS)

Regarding the speed performance of VTCMOS circuits, the gate delay is the vital factor. In CMOS digital circuits, the gate delayed time (t_{pd}) is specified by:

$$Tpd \propto \frac{CVcc}{(Vcc-Vt)^{\alpha}} \tag{4.1}$$

α- power law model.

The operating principle of VTCMOS is that its threshold voltage (V_{th}) is controlled by the applied substrate bias ($-|V_{bs}|$), leading to lower standby off- current or higher active on-current. The V_{th} shift is given by: ΔV_{th} =

$\gamma | V_{bs} |$ where γ is the body effect factor [7]. VTCMOS technique for inverter is shown in figure 4.2 Variable Threshold CMOS or (VTCMOS) is another technique that has been developed to reduce standby leakage currents, relatively, than applying multiple threshold voltage options. VTCMOS relies on a triple well process where the device Vth is dynamically adjusted by biasing the body terminal.

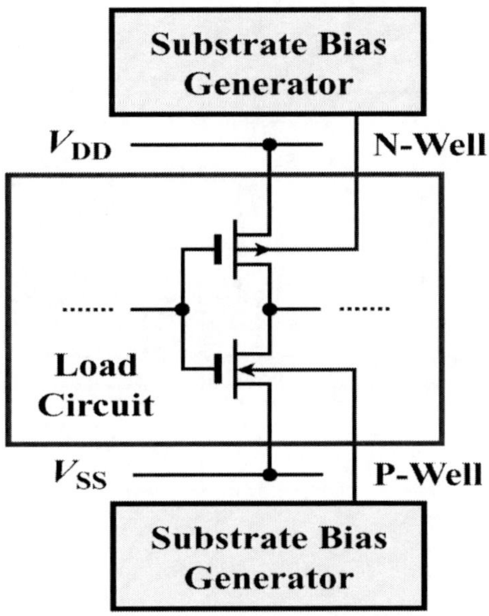

Figure 4.2. VTCMOS.

4.2.3. Dual-Threshold CMOS

For a logic circuit, a higher threshold voltage can be assigned to some transistors in non-critical paths so as to decrease the leakage current, whereas the performance is maintained due to the use of low threshold transistors in the critical path(s). Therefore, no additional leakage control transistors are required, and both high performance and low power can be achieved simultaneously. Dual threshold technique is good for leakage power reduction during both standby and active modes without delay and area overhead. Figure 4.3 depicts the path distribution of dual Vth and single CMOS for a 4:1 MUX.

The static technique approach does not involve the run time decision of this operation for controlling the leakage power. Most leakage reduction

techniques are the dual sub-threshold voltage in these approaches. In the static power reduction, logic level approach has been used for wide form in the custom VLSI design for reducing the leakage power. In the dual-V_t, implementation reduces both the static and the dynamic. The active leakage current and the standby leakage in the CMOS logic VLSI design, in this case, does not affect the performance of the CMOS logic design or it is not degraded in a dual-V_t implementation of the custom design of CMOS VLSI.

There is a lot of work being done, targeted at reducing the leakage power of today's device issues. The reduction techniques can be categorized in two forms i.e. the reduction of leakage in the logic, routing and both. The technique that uses the reduction technique is based on the property of the leakage power consumption in the CMOS logic circuits that is dependent on the state of its inputs and the use of the various statistics' signal to altar the inputs of the state in order to reduce the leakage power consumption in such a way that the is functionality of the circuit does not change [8]. That explains that such multiplexers and the pass transistor logic should have the inputs and outputs of the logic level '1' to reduce the leakage of the circuit in active mode. When the gate leakage of the transistor is of turned OFF, the pass transistor is more than the logic '0', as compared to the case when the logic '1' is passed through it in high state. This technique maximizes the time that is spent on the signal on logic level '1'.

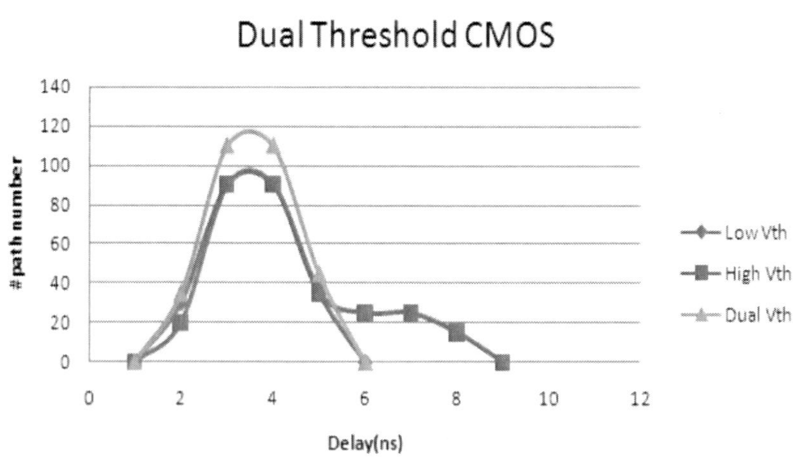

Figure 4.3. Path Distribution of Dual Vth.

4.3. Renowned Techniques That Decrease Leakage Power

They have two well-known techniques that reduce leakage power. Initially, multi threshold-voltage CMOS (MTCMOS) is used [9]. It proficiently decreases the leakage power during the use of high Vt MOSFET switches by disconnecting the power supply. On the other hand, usage of this technique has many drawbacks, both memories and flip-flops based on this technique cannot retain data like the information. The advanced method is to use a variable threshold voltage CMOS (VTCMOS) [10] that reduces leakage power by raising the substrate-biases. Some critical problems are also faced by this technique, through the substrate-bias supply circuits like a large area penalty and a large power penalty. To determine the previously mentioned drawbacks, a small leakage current decrease circuit called a Self-controllable Voltage Level (SVL) circuit has been developed that not only expansively decreases leakage power, other than also retains data through a stand-by period. For upcoming CMOS devices, the amount of gate leakage current has enhanced progressively and is credible to become equivalent or still better to the sub-threshold leakage [11]. In SRAM cells of the abundant techniques which have been proposed to reduce sub-threshold leakage [12]-[13], in active mode a self-controllable switch can be applied [14] which allows full supply voltage to be applied and decreased supply voltage appears to be mainly capable for reducing gate leakage currents as well [15].

4.3.1. SVL Technique

The standby leakage power and the leakage current reduction circuit are known as Self controlable Voltage Level (SVL). The SVL circuit consists of an Upper SVL (U-SVL) circuit and a Lower SVL (L-SVL) circuit, where 2:1 MUX has been used as the load circuit.

There are three types of Self-controllable Voltage Level(SVL):

4.3.1.1. Upper SVL Circuit

Upper SVL consists of a single pmos1v (PM13) and m- nmos1v (NM12 and NM14) connected in series. The 'on- pmos1v' connects a power supply (V_{DD}) and the load circuit is in active mode on demand, and 'on nmos1v' connects a power supply (V_{DD}) and the load circuit is in standby mode. The

schematic of Upper SVL (U-SVL) is shown in figure 4.4. While gate voltage (VG) of the stand-by load circuit is kept at '0', the pmos1v is turned ON while the nmos1v is turned OFF. When control signal (CLB) turns ON nmos1v (NM12) and turns OFF pmos1v, m-nmos1v V_{DD} is supplied circuit, which is called load circuit. Therefore, a drain voltage (V_D) of the 'off nmos1v', a drain-to-source voltage (Vdsn), can be given by the expression:

$$Vdsn = V_{DD} - mv \qquad (4.2)$$

Where v is a voltage drop of the single nmos1v and Vdsn can be changed by varying m or v (or both). Reducing Vdsn by rising mv will enhance the barrier altitude of the 'off nmos1v' i.e., it will decrease the Drain Induced-Barrier-Lowering (DIBL) achieve and, as an effect, it enhance Vthn. This results in a decrease in the sub- threshold current of the nmos1v (Istn) i.e., the leakage current through the load circuit decreases.

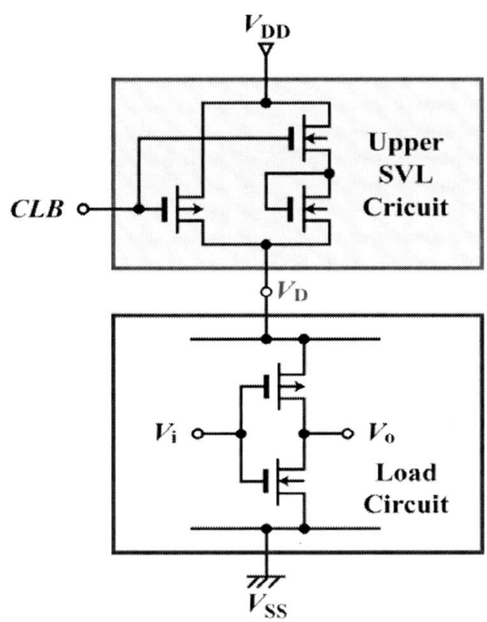

Figure 4.4. Upper SVL Circuit.

4.3.1.2. Lower SVL Circuit

The Lower SVL circuit, which incorporates a single nmos1v (NM3) and m-pmos1v (PM1 and PM3) connected in series, is positioned between a ground-connection (V_{SS}) and the load circuit. The L-SVL circuit not only provides ground connection V_{SS} to the active load circuit through the 'on nmos1v' but also supplies V_{SS} to the standby load circuit through the use of the 'on pmos1v'. The schematic of 2:1 load with L-SVL circuit is put up in figure 4.5. A negative control signal (CL) turns ON pmos1v (PM1) and turns OFF nmos1v so that V_{SS} is supplied to the stand-by load circuit with VG of '0' (i.e., mv) during m-pmos1v. Thus, according to equation (4.3), reduced Vdsn reduces Istn. Furthermore, source voltage (VS) is increased by mv, so the substrate bias, also called back-gate bias (Vsub), is given by expression:

$$Vsub = -\,mv \tag{4.3}$$

Both the reduction in the DIBL effect and the increase in the Back-Gate Bias (BGB) effect lead to further increase in Vthn.

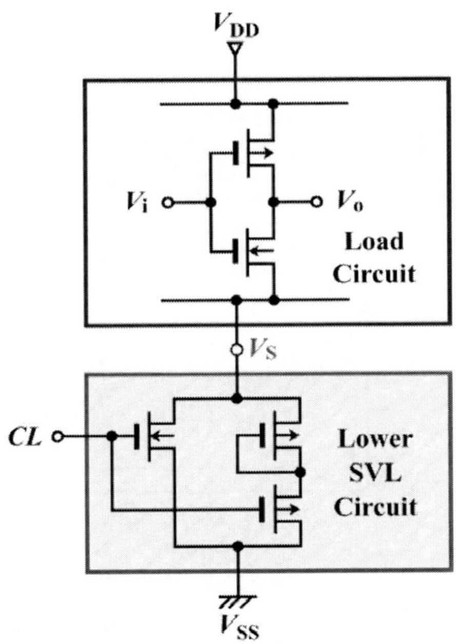

Figure 4.5. Lower SVL Circuit.

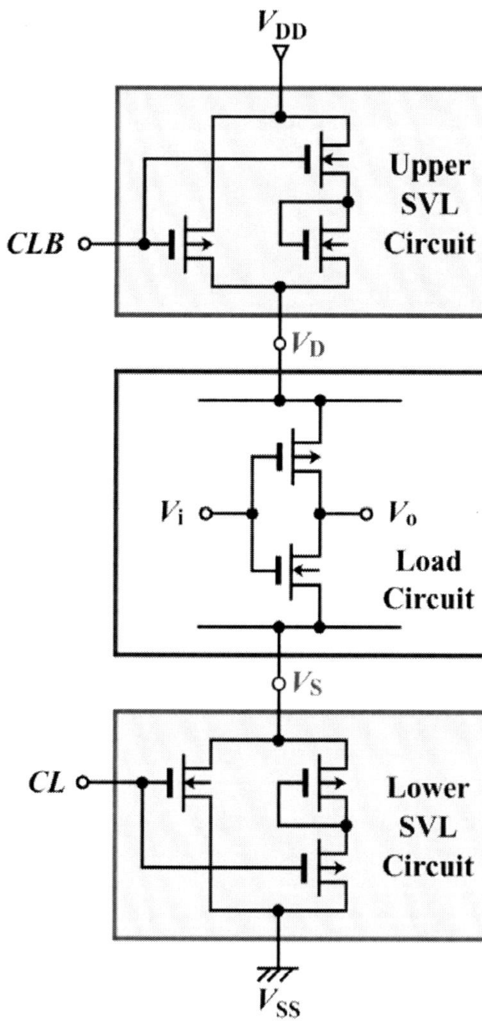

Figure 4.6. SVL Circuit.

4.3.1.3 SVL Circuit

While the load circuit is in the active mode, the built-up SVL circuit supplies the maximum DC voltages (V_{DD} & V_{SS}) to them through U-SVL and L-SVL circuits as both are turned ON simultaneously. Thus the 2:1 MUX/load circuit can operate quickly. On the other hand, when the load circuit is in standby mode, it supplies to some extent lower V_{DD} and relatively higher V_{SS} to them through 'on SVL', so the drain source voltages of the 'off MOSFET'

in the standby load circuit decreases and Vsub increases. Thus, Vth increases and consequently, sub-threshold current decreases, so (standby power) Pst is reduced, while data is retained and noise immunity is high. In the 'cut-off' MOSFETs the increase in VBGs will increase the Vts. The schematic of 2:1 load with SVL circuit is exhibited in figure 4.6. The DIBL effect on nmos1v in the stand-by load circuit, incorporating the SVL circuit, is further decreased, since Vdsn in this case can be expressed as:

$$Vdsn = V_{DD} - 2mv \quad (4.4)$$

The BGB effect due to Vsub which is given in equation (2) also occurs.

4.4. Body Bias Technique

A body bias technique is proposed in [16]. This technique, applied on different topologies, uses more area but works at lower supply voltages. Therefore, it decreases the power consumption [17]. Recent research has shown that minimum energy is typically achieved when it enters the sub-threshold region [18]. Sub- threshold circuits have been shown to be well-designed below 200 mV [19] and with energy consumption from the category of pico Joules per instruction [20]. The challenges of ultra-low-voltage memory design in the recent work have also been explored [21] in previous work and supposed to have reduced noise margins at low voltage [22], wherein gates with large fan-ins have been shown, [23]. Using a robust latch-based memory with a MUX-based read-out structure, the data memory, instruction memory and register file were realized [24]. It was shown by increasing the gate sizes along critical paths due to the timing dependence of leakage energy that the energy of a sub-threshold circuit can be reduced [25]. Since simulations show that body current are several orders of magnitude smaller than sub-threshold current for the body bias range it was explored the reported energy numbers only include current drawn from the generator [26]. Benefits of body biasing, without considering the costs of body 886 bias regulations, we also quantify the energy and delay [27]. Past work has discovered the proficient production and instruction of both and body bias [28].

4.4.1. Body Bias for Variability Control

Process variability has the potential to be a crippling problem in sub-threshold circuits. Variation is typically classified as random within-die, correlated within-die, or die-to-die. Random within-die variations in gate length are becoming particularly problematic in scaled devices [29]. For example, random variations induced by Random Dopant Fluctuations (RDFs) have an inverse dependence on the gate area, giving 3 variations as high as 58 mV at the 65 nm technology node [30]. Given an inverse sub-threshold slope of 85 mV/dec, such large variation can yield 3 / current fluctuations as elevated as 4.8 the insignificant. Interrelated variations are also a major problem at low voltage. In this research, we group interrelated within-die and die-to-die variations mutually under the name global variation due to the small size of our sub-threshold processor. Global variations can be either static process variations (e.g., gate length or gate oxide thickness) or temporal variations (e.g., temperature) [31]. Sub-threshold operation is dominated by an exponential dependence, so global variations are due to doping fluctuations or those induced by oxide thickness, gate length and temperature variations are main cause of concern.

4.4.2. Body Bias in Sub-Threshold Circuit

Body biasing, which effectively skews, has been suggested for global compensation in the past. The authors [32] designed a multiply-accumulate unit that used adaptive supply voltage and body bias to minimize power in super-threshold circuits. In the sub-threshold regime, Body biasing is mainly an efficient technique due to the exponential addiction of sub-threshold current on body bias. The use of Body biasing in sub-threshold circuits was briefly explored in [33], but little attention was given to how the body bias should be selected and only limited measurements were presented. The authors of [34] showed that correct operation can be achieved with as low as 100 mV by tuning body biases to match P-type FET and N-type FET leakages. In this study, we extend these early studies to develop a comprehensive body-biasing strategy that accounts for the unique sensitivities of sub-threshold circuits. We also present detailed measurements of 20 measured dies to confirm the observed trends.

4.4.3. Forward Body-Bias Method

MOSFET is traditionally treated as a three-terminal device in a circuit design as shown in figure. 4.7, where source and substrate terminal are shorted. It implies that the MOSFET operates as a zero-biased threshold device. Further, when it used as a four-terminal device, a forward or reverse bias can be applied between the source and substrate of a MOSFET. The MOSFET threshold voltage is improved when it is used as the Reverse Body-Bias. Under the Forward Body-Bias, the MOSFET threshold voltage is reduced. The relationship between the threshold voltage and the body-bias is depicted in figure 4.8 (a) shows a standard CMOS inverter with shorted body-source transistors. Figure 4.8 (b) shows a CMOS inverter under Forward Body-Bias conditions where the MOSFET is used as a four terminal device.

The back gate forward body-bias method is shown in figure 4.8 (b) which is attuned with CMOS processes. Without any technology adaptation the threshold voltage of MOS transistors can be decreased electrically.

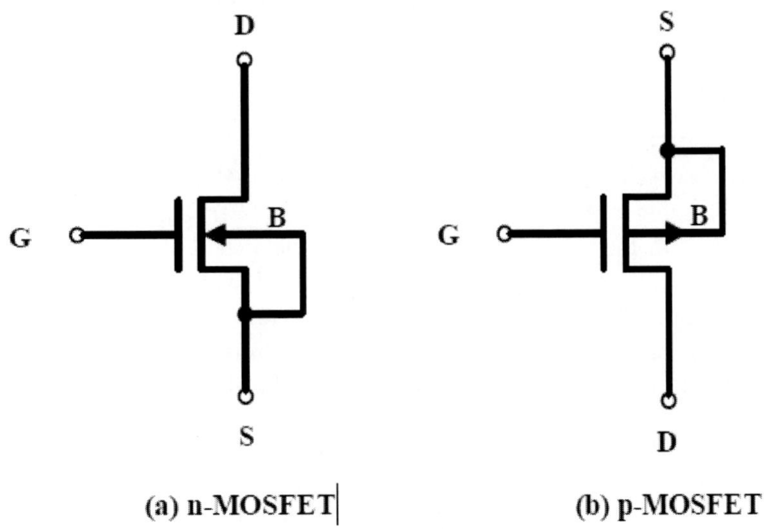

Figure 4.7. MOSFET as a three-terminal device (a) n-MOSFET (b) p-MOSFET. Here, S, G, D and B, respectively indicate source, gate, drain and substrate.

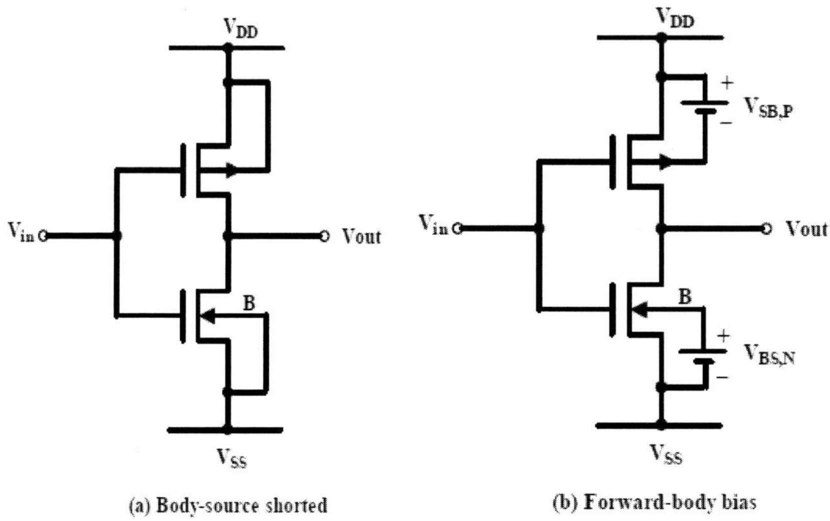

Figure 4.8. CMOS inverter under (a) Body-Source Short and (b) Forward-Body Bias conditions. The VBS, N is the Forward Body-Bias across the n-MOSFET and VSB, P is the Forward Body-Bias across the p-MOSFET.

REFERENCES

[1] K. Roy, S. Mukhopadhyay, H. Mahmoodi-Meimand, "Leakage current mechanisms and Leakage reduction techniques in deep-Submicrometer CMOS circuits" leakage," *Proceeding of IEEE,* vol. 91, no.1 pp. 305-327, Feb 2003.

[2] J. Kao, S. Narendra, and A. Chandrakasan, "Subthreshold leakage modeling and reduction techniques," *Proceeding of IEEE/ACM International Conference on Computer-Aided Design (ICCAD),* pp. 141-148, 2002.

[3] S. Mutah, T. Douseki, Y. Matsuya, T. Aoki, S. Shigematsu, J. Yamada, "1-V power supply high speed digital circuit technology with multi-threshold voltage CMOS," *IEEE Journal of Solid State Circuits*, vol. 30, no. 8, pp. 847-854, 1995.

[4] M. Anis, M. Mahmoud, M. Elmasry, S. Areibi, "Dynamic and leakage power reduction in MTCMOS circuits using an automated efficient gate clustering technique," *Proceeding of IEEE 39th Design Automation Conference (DAC),* pp. 480-485, 2002.

[5] J. Kao, A. Chandrakasan, D. Antoniadis, “Transistor sizing issues and tools for Multi-threshold CMOS technology,” *Proceeding of IEEE 34th Design Automation Conference (DAC),* pp. 409-414, 1997.

[6] S. Mutoh, T. Douseki, Y. Matsuya, T. Aoki, S. Shigematsu, and J. Yamada, “1-V power supply high-speed digital circuit technology with multi-threshold voltage CMOS,” *IEEE J. Solid-State Circuits,* vol. 30, pp. 847–854, Aug. 1995.

[7] T. Hiramoto and M. Takamiya., “Low power and low voltage MOSFETs with variable threshold voltage controlled by back-bias”, *IEIEC Trans. Electron.,* vol. E83- C, No.2, 2000.

[8] Jason H. Anderson, F.N. Najm, and T. Tuan., “Active leakage power optimization for FPGAs,” FPGA, *IEEE Transactions On Computer-Aided Design of Integrated Circuits and Systems,* vol. 25, no. 3, pp. 423-437, March 2006.

[9] J. Rodrigues, O. C. Akgun, and V. Owall, “A <1 pJ sub-VT cardiac event detector in 65 nm LL-HVT CMOS,” in Proc. VLSI-SoC, June 2010.

[10] O. Akgun, J. Rodrigues, Y. Leblebici, and V. Owall, “High-level energy estimation in the sub-VT domain: Simulation and measurement of a cardiac event detector,” *IEEE Trans. Biomed. Circuits Syst.,* accepted for publication.

[11] Mizuno H and Nagano T (2010) Driving source-line cell architecture for sub-1-V high-speed low-voltage applications. *IEEE J. Solid State Circuits.* 31(4), 552-557.

[12] Mutoh S et al. (2008) A 1V multi-threshold voltage CMOS DSP with an efficient power management technique for mobile phone application. Digest of technical papers, *IEEE Int. solid-state circuits conf.* (ISSCC’96), FA 10.4, pp168-169.

[13] Kuroda T et al. (2009) A 0.9-V, 150-MHz, 10-mW, 4-mm2 2-D discrete cosine transform core processor with variable threshold-voltage (VT) scheme. *IEEE J. Solid-State Circuits.* 31(11), 1770-1779.

[14] L. Chang et al., “Stable SRAM Cell Design for the 32nm Node and Beyond,” *Symp. VLSI Tech. Dig.,* pp. 292-293, Jun., 2005.

[15] Rafik S. Guindi, Farid N. Najm, Design Techniques for Gate-Leakage Reduction in CMOS Circuits, *Proceedings of the Fourth International Symposium on Quality Electronic Design,* p.61, March 24-26, 2003.

[16] M. E. Hwang and K. Roy, “*A 135mV 0.13μW Process Tolerant 6T Sub threshold DTMOS SRAM in 90nmTechnology*” CICC, pp.1-4, 2008.

[17] B. Zhai, D. Blaauw, D. Sylvester, and K. Flautner, "Theoretical and practical limits of dynamic voltage scaling," in *Proc. Design Automation Conf.,* 2004, pp. 868–873.

[18] B. Calhoun and A. Chandrakasan, "Characterizing and modeling minimum energy operation for sub threshold circuits," in *Proc. IEEE Int. Symp. Low Power Electronics and Design (ISLPED),* 2004, pp. 90–95.

[19] A. Wang and A. Chandrakasan, "A 180 mV FFT processor using sub threshold circuit techniques," in *IEEE Int. Solid-State Circuits Conf. (ISSCC) Dig. Tech. Papers,* 2004, pp. 292–293.

[20] B. Zhai, L. Nazhandali, J. Olson, A. Reeves, M. Minuth, R. Helfand, S. Pant, D. Blaauw, and T. Austin, "A 2.60 pJ/inst sub threshold sensor processor for optimal energy efficiency," in *Symp. VLSI Circuits Dig. Tech. Papers,* 2006, pp. 154–155.

[21] L. Chang, Y. Nakamura, R. Montoye, J. Sawada, A. Martin, K. Kinishita, F. Gebara, K. Agarwal, D. Acharyya, W. Haensch, K. Hosokawa, and D. Jamsek, "A 5.3 GHz 8T-SRAM with operation down to 0.41 V in 65 nm CMOS," in *Symp. VLSI Circuits Dig. Tech. Papers,* 2007, pp. 252–253.

[22] B. Zhai, D. Blaauw, D. Sylvester, and S. Hanson, "A sub-200 mV 6T SRAM in 130 nm CMOS," in *IEEE Int. Solid-State Circuits Conf. (ISSCC) Dig. Tech. Papers,* 2007, pp. 332–333.

[23] S. Hanson, B. Zhai, K. Bernstein, D. Blaauw, A. Bryant, L. Chang, K. Das, W. Haensch, E. Nowak, and D. Sylvester, "Ultra-low voltage minimum energy CMOS," *IBM J. Res. Devel.,* vol. 50, no. 4/5, pp. 469–490, 2006.

[24] J. Chen, L. Clark, and Y. Cao, "Robust design of high fan-in/out sub threshold circuits," in *Proc. Int. Conf. Computer Design,* 2005, pp. 405–410.

[25] S. Hanson, D. Sylvester, and D. Blaauw, "A new technique for jointly optimizing gate sizing and supply voltage in ultra-low energy circuits," in Proc. IEEE Int. Symp. *Low Power Electronics and Design (ISLPED),* 2006, pp. 338–341.

[26] Y. Ramadass and A. Chandrakasan, "Minimum energy tracking loop with embedded DC-DC converter delivering voltages down to 250 mV in 65 nm CMOS," in *IEEE Int. Solid-State Circuits Conf.* (ISSCC) Dig. Tech. Papers, 2007, pp. 64–65.

[27] C. H. Kim, K. Soeleman, and K. Roy, “Ultra-low-power DLMS adaptive filter for hearing aid applications,” *IEEE Trans. Very Large Scale Integr. (VLSI) Syst.,* vol. 11, no. 6, pp. 1058–1067, Jun. 2003.

[28] A. Bryant, J. Brown, P. Cottrell, M. Ketchen, J. Ellis-Monaghan, andE. Nowak, “Low-power CMOS at Vdd = 4kT/q,” in *Proc. Device Res. Conf.,* 2001, pp. 22–23.

[29] F. Moradi, D. Wisland, H. Mahmoodi, Y. Berg and T. Cao, "New SRAM Design Using Body Bias technique for ultra low power application," *11th Int'l Symposium on Quality Electronic Design,* pp. 468-471, 2010.

[30] Jan M Rabaey, Anantha Chandrakasan, Borivoje Nikolic, *“Digital Integrated Circuits: A Design perspective,”* 2nd Edition, Prentice-Hall, Inc.

[31] J.M. Rabaey, A. Chandrakasan, B. Nikolic, *Digital Integrated Circuits – A Design Perspective,* Pearson Educational publishers-2nd edition 2008.

[32] Neil H.E. Weste, David Harris and Ayan Banerjee, *CMOS VLSI Design, A circuits and system perspective,*3rd Edition, Pearson Education, 2005.

[33] J. Rodrigues, O. C. Akgun, and V. Owall, “A <1 pJ sub-VT cardiac event detector in 65 nm LL-HVT CMOS,” in *Proc. VLSI-SoC,* June 2010.

[34] O. Akgun, J. Rodrigues, Y. Leblebici, and V. Owall, “High-level energy estimation in the sub-VT domain: Simulation and measurement of a cardiac event detector,” *IEEE Trans. Biomed. Circuits Syst.,* accepted for publication.

Chapter 5

Impact of GDI and Finfet on Multiplexer

5.1. Objectives of the Chapter

The goal of this chapter is to reduce power and area of 2:1 MUX while maintaining the viable performance. The diverse configurations are designed using different topologies of 2:1 MUX such as topology of MUX was designed using CMOS, transmission gate and pass transistor using FINFET(Fin shaped Field Effect Transistor). The mobility was improved in devices with taller fins due to enhancement tensile stress. To develop performance and reduce power consumption, DG FINFET can be proficiently used. In non-critical paths, sovereign gate control can be used to join mutually parallel transistors.

In this chapter design of proposed reversible logic multiplexer with garbage input and output with Gate -Diffusion -Input (GDI cell) technique are used in circuit realization, in which the low power and area design in this dynamic constituent of power are condensed, In GDI cell, PMOS is not linked to supply voltage VDD and NMOS is not linked to the GND. GDI circuits present some major issues of enhanced hazard tolerance and low voltage operation of reversible logic circuits. In this section, it proposes a new application of GDI (Gate-Diffusion Input) circuits to Reversible logic multiplexer with its Garbage input and output. The novel proposed design technique will consume less power than the other conventional gate. Reversible logic circuit has shown less power dissipation in the recent years. This GDI cell technique decreases the power of the circuit, delay, Power-

Delay Product (PDP) and it also compacts frequency. The device scaling is partial as the power dissipation is more optimized in terms of delay, frequency jitter, bandwidth power supply, frequency and duty cycle of the signal and also establishes the noise of the circuit. The reversible logic design, GDI, is efficient in lowerd Delay, low power and low leakage current. The transistor realization of the proposed gates is done by using Virtuoso tool of cadence

5.2. INTRODUCTION

Due to lower frequency rate, the supply voltage may be reduced below the threshold voltage for wireless sensor network applications. This operation is referred as sub-threshold design that uses the sub-threshold current as constrain current to estimate the inputs [1]. Different 2:1 MUX topologies have been proposed for reliable sub-threshold applications.

In silicon n-channel Field-Effect Transistors (n-FETs) Silicon-Carbon (Si: C) Source/Drain (S/D) stressors may be adopted for attracting the electron mobility and drive current [2]. These values are extensively higher than the doping extracted by electrical categorization, signifying doping loss in fins and/or partial activation [3]. Doped-channel FINFETs are appropriate for system-on-chip applications which require various threshold voltages on the same die. For an unusual device structure and for replacing the planer CMOS device structure, FINFET technology is one of the most proficient candidates [4]. In the operation of planar MOSFET, scaling down of CMOS technology leads to severe short channel effects. However, increase in short channel effects will lead to degradation in the performance. Another device structure is essential, to attain superior control over the gate. They have an intrinsic force against the SCE. Therefore, many studies have focused on the FINFET SRAM cells to overcome the rapid decrease in the conventional bulk SRAM performance [5]. The effect of FINFET variability on the 2:1 MUX has been also studied [6]. In sub-22 nm CMOS technology nodes, due to their superior electrostatic integrity as compared to the conventional planar bulk MOSFET, three-dimensional transistor structures such as double-gate FINFET and tri-gate FET are slated for adoption [7]. Even when the fin width is reduced to ~4 nm to enable gate length (Lg) scaling down to 10 nm, recent experimental results show that the FINFET performs well [8]. The process simulator within the Sentaurus technology computer-aided-design software suite [9], which uses the finite-element method, was used to perform 3-D simulations of stress within FINFETs with (100) top and (110) sidewall surfaces and channel

direction. Since the bulk-silicon substrate provides a pattern for epitaxial growth, so that the entire S/D regions are anxious for bulk FINFETs, the fin S/D regions are implicit to be etched away earlier to the discriminating epitaxial, [10] which is a gate-last (i.e., replacement metal gate) process flow, in which a dummy gate is formed earlier to the S/D epitaxial and then replaced by the metal gate [11]. The 1-D stress profiles are extracted at 1 nm beneath the fin sidewall surfaces (corresponding to inversion-layer position [12]) and averaged diagonally to the fin height. For short-channel Si MOSFETs, carrier mobility is a good pointer of device performance [13]. To expect the stress-induced carrier mobility improvement in a physically based approach (e.g., to version for the nonlinear piezo-resistance effect in strained MOSFET channels), a self-consistent [14] Poisson–Schrödinger equation solver was developed in the previous work which is used to compute the silicon sub-band structure and efficient carrier mobility values for FINFET structures [15]. Only the conduction along the fin sidewall surfaces is measured, to be pertinent for insistently scaled FINFETs [16].

Reversible logic is a very prospective approach of logic synthesis for power reduction in future computing technologies [17]. In a reversible logic circuit, the number of inputs and outputs are same and there is one-to-one mapping between input and output values. Reversible circuits are constructed using reversible gates. The recent computers erase a bit of information every time they perform a logical operation. Such logical operations are usually identified as ‘irreversible logic’. Zero energy dissipation would be possible only if the network consists of reversible gates. So that reversibility will become an essential property in future circuit design [18]. According to Landauer [19],

$$E_{bit} \geq E_{SNL} = KBTln2 = 0.017eV \tag{5.1}$$

Where, E_{bit}=the energy required for binary transition, ESNL=Shannon-von Neumann Landauer (SNL) expression

Traditional irreversible hardware computation, inevitably, leads to energy dissipation due to the loss of each one bit of information which dissipates an amount of KTln(2) Joules of energy, where K is the Boltzmann’s constant (1.3807×10^{-23} JK^{-1}) and T is the absolute temperature at which estimation is performed. For high technology circuits the heat that is degenerated is very large, when compared to a loss of one bit information at room temperature. Furthermore, voltage-coded logic signals have energy of $E_{sig}=1/2CV^2$ and this energy is dissipated whenever the node voltage changes in the irreversible

CMOS technology. The technology scaling is to be used to reduce power consumption in which the supply voltage is proportional to the threshold voltage. A new low power design technique is that it solves most of the problems with the help of this technique i.e. Gate-Diffusion-Input (GDI) as discussed in [20]. This technique allows the reduction of power low, low leakage current, propagation delay and Power- Delay Product (PDP). In this chapter, we present a novel reversible logic multiplexer proposed circuit with its garbage input/output. One of the major concern issues is that require a Twin-well CMOS or silicon on insulator process is realizing [21]. It is estimated that reversible logic circuits also helps to secure energy by using charge recovery logic [22]. As the Moore's law continues to hold, the processing power just doubles every 18 months. The present irreversible technologies will dissipate a lot of heat and can reduce the life of the circuit. The reversible logic operations will be unable to find (erase) information and dissipate very less heat. Thus, reversible logic circuits are likely to be in demand in high-speed power circuits. Reversible logic supports both the processes of running forward and backward. This means the reversible computation can generate the inputs from outputs and can stop and go back to any point in the computation operation [23]. Reversible logic circuits are of high interest in low-power CMOS design, nanotechnology, quantum computing and optical computing. The most important application of reversible logic lies in quantum computers [24]. A quantum computer will be viewed as a quantum network (or a family of quantum networks) composed of quantum logic gates; each gate performing an elementary unitary operation on one or more than two state quantum.

5.3. CMOS Based 2:1 MUX

In electronics, a multiplexer is a device that selects one of a number of analog or digital input signals and frontwards the selected input into one signal. Multiplexers are generally used to increase the amount of data that can be sent over the network within a certain amount of time and bandwidth. A Multiplexer is also called the data selector. In Digital circuit design, the selector wire is of digital value. In case of 2:1 multiplexer, a logic value of '0' would attach A to the output while the logic value of '1' would attach B to the output.

A 2:1 multiplexer has a Boolean expression where A and B are two inputs Select line (S) is the selector input and Vout is the output.

$Vout = (A.S_0) + (B.S_1)$

A basic realization of 2-to-1 MUX would need 2 AND gate, an OR gate and a NOT gate. Symbol Diagram of 2:1 MUX is shown in Figure 5.1, Schematic of CMOS based 2:1 MUX is shown in Figure 5. 2. Any logic function in complementary CMOS is realized by NMOS pull-down and PMOS pull-up networks connected between gate, output and power lines [25].

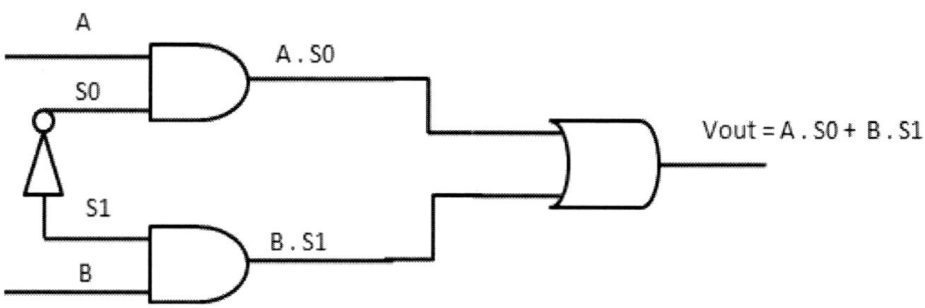

Figure 5.1. Symbol Diagram of 2:1 MUX.

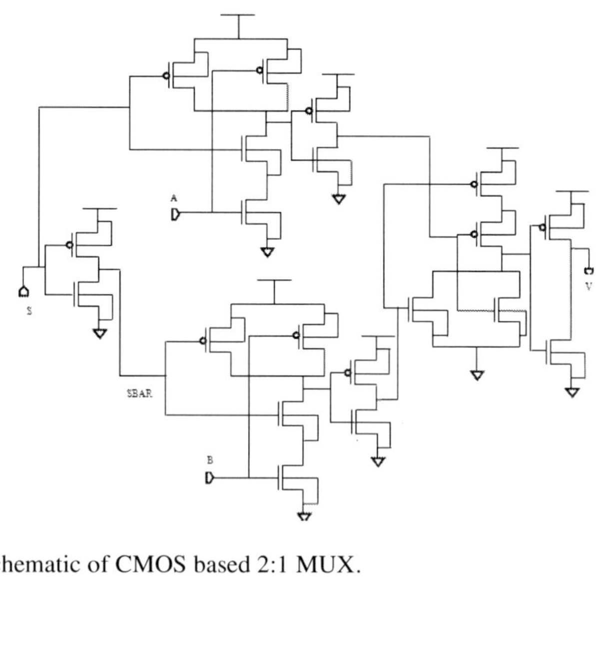

Figure 5.2. Schematic of CMOS based 2:1 MUX.

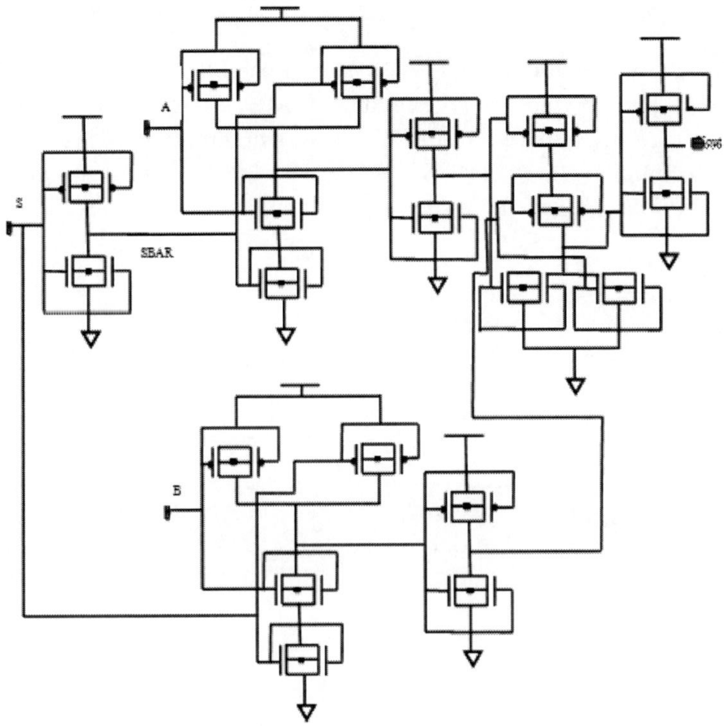

Figure 5.3. Schematic of CMOS based 2:1 MUX using FINFET.

5.3.1. CMOS Based 2:1 MUX Using FINFET

CMOS based 2:1 MUX using FINFET schematic is given figure 5.3. In digital and analog circuit designs, Double Gate devices have been used in a variety of creative ways. A parallel transistor couple consists of two transistors with their source and drain terminals attached together. The second gate is added opposite to the conventional gate in Double-Gate (DG) FINFETS, which have been known for their probable to advance control short channel effects, as well as to control leakage current. The operations of FINFET is predictable as Short Gate (SG) mode with transistor gates joined together, the low-power and optimum power mode where the back gate is attached to a reverse-bias voltage to reduce leakage power and the hybrid mode, the independent gate (IG) mode where sovereign digital signals are used to constrain the two device gates, which employs a collection of low power and

sovereign gate modes. Here self-determining control of front and back gate in DG FINFET can be proficiently used to extend performance and decrease power consumption. In non-critical paths, sovereign gate control can be used to connect together parallel transistors together.

5.4. Transmission Gate Based 2:1 MUX

This is the Transmission Gate based 2:1 MUX structure implemented with very minimum transistors (4 MOS transistors) compared to the CMOS based 2:1 MUX which has 20 CMOS devices. The back to back connected PMOS & NMOS transistors arrangement acts as a switch is the so called 'Transmission Gate'. In Transmission Gate NMOS transistor passes a strong '0', but a weak '1', while PMOS transistor passes a strong '1', but a weak '0'. The CMOS based 2:1 MUX uses NMOS transistor which acts as pull down network and PMOS transistor acts as pull up network, whereas the transmission gate, combines both the properties by placing NMOS transistor in parallel with the PMOS transistor. Two transmission gates are connected as shown under the schematic of Transmission based 2:1 MUX in figure 5.4 to form a MUX structure [26]. Each Transmission Gate acts as an AND switch to replace the AND logic gate which is used in a CMOS Based design of MUX. Hence, the transistor count is reduced to 4, so it occupies less area as compared to CMOS Based MUX. One more change when compared to CMOS Based 2:1 MUX is that there is no supply voltage applied to the circuit. It results in less operating power. It has lower gate delay and the circuit propagates faster than that of the CMOS Based 2:1 MUX. The gate delay as mentioned can be computed as equation (5.1)

$$t_{pd} \propto C_L V_{dd} / I_{ds} \qquad (5.1)$$

Where, t_{pd} = propagation delay,
CL = load Capacitance,
V_{dd} = supply voltage and
I_{DS} = drain saturation current

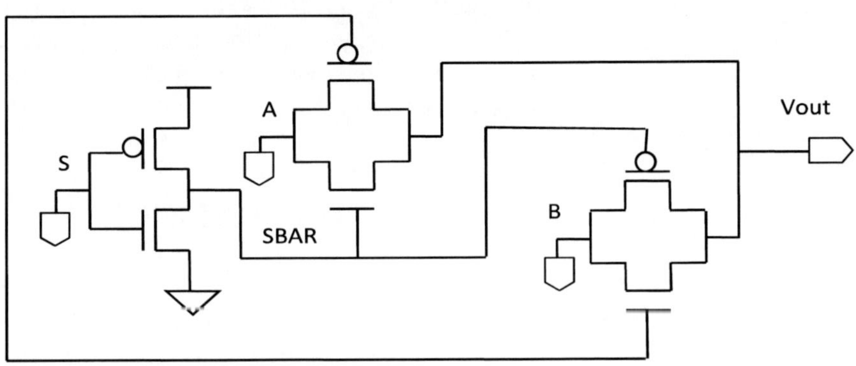

Figure 5.4. Schematic of Transmission Gate 2:1 MUX.

5.4.1. Transmission Gate 2:1 MUX Using FINFET

Transmission Gate Based 2:1 MUX using 6 transistors is displayed are shown in figure 5.4. Transmission Gate based 2:1 MUX using FINFET is illustrated in figure 5.5. A parallel transistor pair consists of two transistors with their source and drain terminals tied together. The second gate is added opposite to the conventional gate in Double-Gate (DG) FINFETS in Transmission gate, which have been predicted to be prospective to superior control short channel effects and also control leakage current.

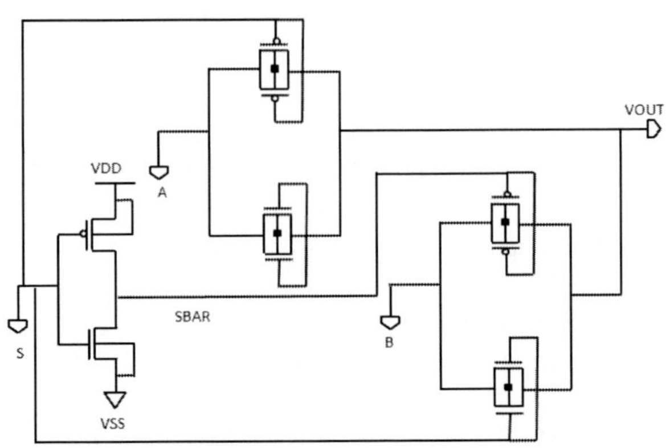

Figure 5.5. Schematic of Transmission gate 2:1 MUX using FINFET.

5.5. Pass Transistor Based 2:1 MUX

The basic difference of pass-transistor based 2:1 MUX compared to the CMOS Based 2:1 MUX is that the source side of the logic transistor networks is connected to some input signals instead of the power lines. The benefit is that it uses one PMOS and one NMOS transistor sufficient to perform the logic operation, which results in a minimum number of transistors (2 MOS transistor) and these two pass transistors, at the input, select which signal to propagate, especially when NMOS networks are used. The Schematic of Pass Transistor Based MUX is shown in figure 5.6. The logic levels will be deteriorated by the pass transistor. The threshold voltage of both pass-transistors should be one and the same for accurate operation. However, the threshold voltage drop (Vout= Vdd-Vtn) through the NMOS transistors while passing a logic '1' makes move backward and forward (or level) return at the gate outputs, necessary, in order to avoid constant currents at the consequent output inverters or logic gates. Adjusting the threshold voltages (i.e., Vtp>Vtn) as a solution at the process technology level is usually not sufficient for other reasons. Among all these Topologies of 2:1 Multiplexer, it is the optimal.

It is efficient and compared to other topologies occupies less area. Finally, pass-transistor logic with move backward and forward re-establishment circuitry is sensitive to voltage scaling and transistor sizing with respect to circuit robustness (reduced low power and area occupied), i.e., efficient or reliable operation of logic gates is not essentially certain at low voltages or small transistor sizes. There are several pass-transistor logic styles like Complementary Pass Transistor Logic (CPL) and Energy Economized Pass Transistor Logic (EEPL) [27].

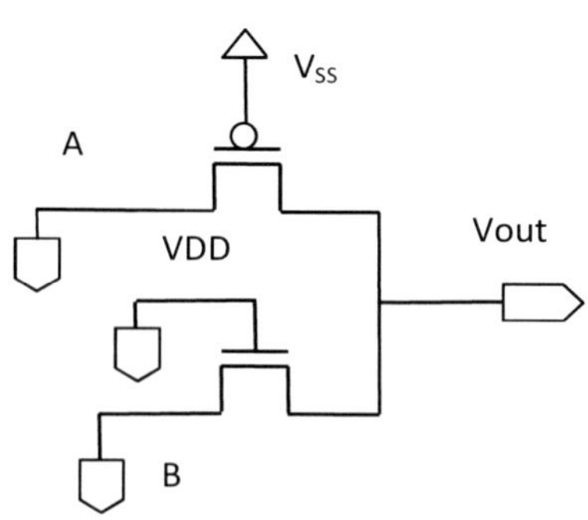

Figure 5.6. Schematic of Pass Transistor 2:1 MUX.

5.5.1. Pass Transistor 2:1 MUX Using FINFET

Pass Transistor Based 2:1 MUX using 2 Pass transistors is exhibited in figure 5.7. Due to its base material, the uninterrupted down in scaling of bulk CMOS creates primary issues. The crucial obstacles to the scaling of bulk CMOS to 45nm gate lengths include short channel effects, gate-dielectric leakage and device to device variations. But FINFET based designs offer a superior control over short channel effects, low leakage and better yield [28] in 45nm helps to overcome the obstacles in scaling.

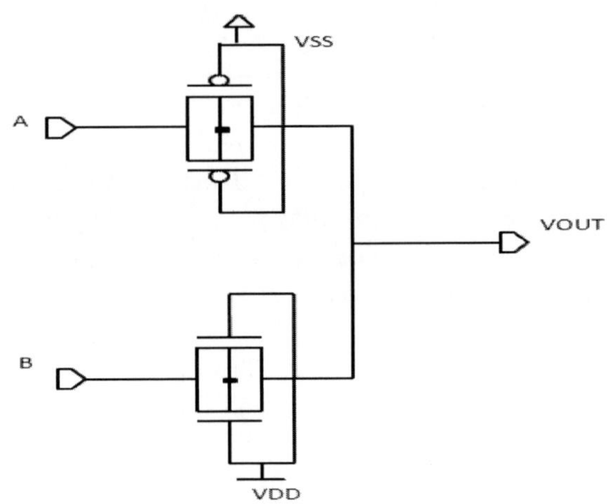

Figure 5.7. Schematic of Pass Transistor 2:1 MUX using FINFET.

5.6. Basic Concept of Gate Diffusion Input (GDI) Cell

The basic concept of GDI cell technique is shown in figure 5.8. The PMOS in GDI cell is not connected to the supply VDD and on the other hand, NMOS in GDI cell uncorrelated to GND. This characteristic specifies that the GDI cell technique has two extra input pins to use which makes the design more compatible as no input increasing the digital circuits. It must be remarked that not all the functions are possible in standard P-well process but the most successfully design process is the CMOS process. One of the major problems of Gate-diffusion input technique is that it requires a special CMOS process whereas the GDI schemes require twin-well CMOS or Silicon-On-

Insulator (SOI) process to implement the design which makes it more expensive than the other standard cells [29]. One of the common basic problems is occurring in GDI cell based logic design which is the low swing of output signals because of the threshold voltage drop across the single channel pass transistor. A simple change is to be brought in the input configuration of GDI cell based on Boolean function that is different every function. [30]. The reverse logic multiplexer with garbage input/output is more and the power dissipation in the circuits is less in comparison to the other CMOS logic reversible designs [31]. GDI Technique enables simpler reversible gates; transistor count is less and lower power dissipation in many implementations, as compared with other CMOS design techniques circuits in VLSI Design. The new device can have the basic wide spectrum of implementation in a wide variety of nanotechnology circuits.

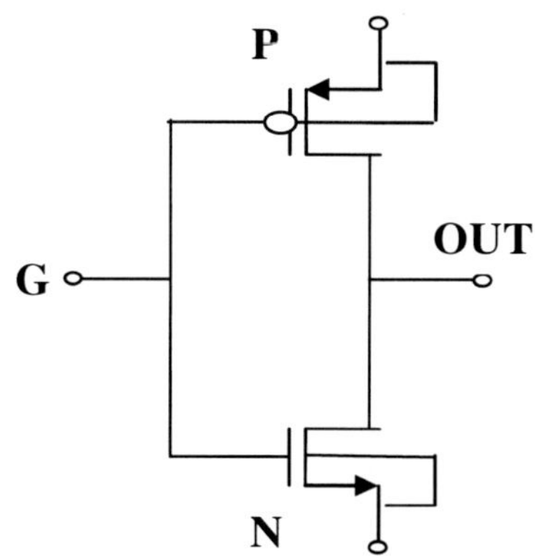

Figure 5.8. Basic Gate-Diffusion Input Cell.

The future work of reversible logic multiplexer is modeling and simulation of the new Nano devices [32]. Also the basic GDI cell based functional is figured in table 5.1. The functions of F1 and F2 are the absolute necessities of logic families. F1 is the GDI function and its functionality can be realized for the standard P-well CMOS process, because the bulk of any NMOS is equally or constantly biased [33]. The GDI cell technique is effective for the low power design operation and it is worthy to understand low delay, high PDP and high frequency response of the logical circuit.

Table 5.1. Truth Table of the Basic GDI Cell (Gate Diffusion Input)

N	P	G	OUT	Function
‘0’	B	A	$\bar{A}B$	F1
B	‘1’	A	$\bar{A}$+B	F2
‘1’	B	A	A+B	OR
B	‘0’	A	AB	AND
C	B	A	$\bar{A}$B+AC	MUX
‘0’	‘1’	A	$\bar{A}$	NOT

5.6.1. Basic Reversible Logic Gate

A reversible logic gate is n × k logic device where n and k is the number of inputs and outputs with one-to-one mapping, i.e. and should be equal to k (n = k). A gate is said to be reversible if the input vector can be uniquely recovered from the output vector and there is a one-to-one correspondence between its input and output logic [34]. A set of reversible logic gates is needed to design reversible logic circuits. An n × k reversible logic gate can be represented as:

$$I_V = (I_1, I_2, I_3 \dots \dots \dots I_n) \quad (5.2)$$

$$O_V = (O_1, O_2, O_3 \dots \dots \dots O_k) \quad (5.3)$$

Where, I_v and O_v are input and output vectors. Consider the following issues to perform the synthesis of reversible gates and obtain optimization. Any reversible circuit realizes only the function that is reversible. The number of outputs in a reverse gate or circuit are the same as the number of inputs, and generally used traditional NOT gate is the only reverse gate. Every Reversible gate has a cost associated with it, called Quantum cost. The simplest reversible gate is NOT gate and is a 1×1 gate [35].

The Reversible gate 2×2 with quantum cost of the reversible gate having mapping input (A,B) to output (P = A, Q = AΘB) is displayed in figure 5.9 that the combinational circuit diagrams of the NOT gate and CNOT gate [36].

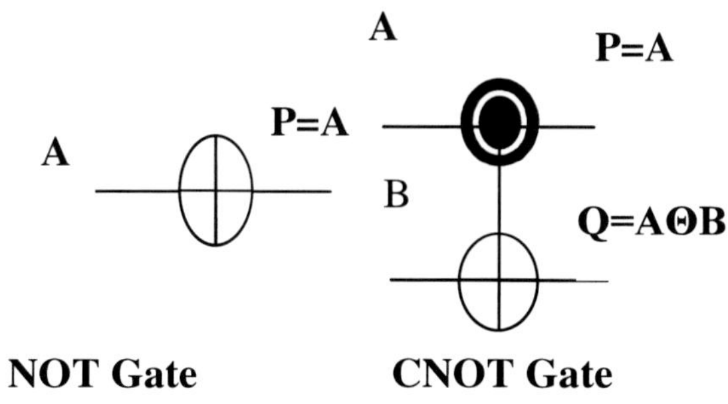

Figure 5.9. Combinational Circuit Diagram of 1x1 NOT Gate and CNOT Gate.

The controlled NOT (CNOT) gate is an example of a 2×2 gate. Many of the available gates are 3×3 Reversible gates such as TR, FG, PG and TG gates. The Quantum Cost of 1×1 Reversible gates is zero, and the Quantum Cost of 2×2 Reversible gates is one. Any reversible gate is optimized by using 1×1 NOT gates (inverter) and 2×2 (Feynman gate) Reversible gates, such as V, V^+ (V is the square root of NOT gate and V^+ is its Hermitian) and FG gate that is also known as CNOT gate. The properties of V and V^+ Quantum Gates are given in the following equation [37].

$$V \times V = \text{NOT} \tag{5.4}$$

$$V \times V^+ = V^+ \times V = I \tag{5.5}$$

$$V^+ \times V^+ = \text{NOT} \tag{5.6}$$

The Quantum Cost of a Reversible gate is calculated by counting the number of V, V^+ and CNOT gates. One of the most important attributes of a reverse gate is its garbage output, i.e. each and all inputs of the gate that are not used as input to another gate or as a primary output is called garbage output. Garbage input/output is the required for every reversible gate logic system.

The constants that are used as the input line are either set to Zero (0) or One (1) in the circuit input side. Gate count is the number of reversible gates used to realize the function. The hardware complexity refers to the number of basic gates (NOT, AND, and EXOR gate) used easily to synthesize more given functions [38]. The delay can be calculated in reversible logic based on

the critical path. It is the longest path that gives the desired output of the system [39]. There exist many reversible gates in the literature. The proposed 3×3 reversible gate with two of its output working as 2:1 multiplexer, the gate to used a design of reversible gate circuit with its garbage inputs and outputs, also design a Reversible half-adder and half- subtractor circuits. Among them Feynman gate (FG), Peres gate (PG), Toffoli gate (TG), Fredkin gate (FRG) and Khan Gate (NG) are very common [40]. The quantum realizations of all these gates are not available in literature. Only PG, FG, TG, and FRG have been realized in Nanotechnology. A reversible logic circuit should have the following features.

- Least number of reversible gates
- Least number of constant inputs
- Least number of garbage outputs
- Length of cascading gates should be minimum
- Similar number of outputs so as to facilitate with that of inputs
- Hardware complication i.e. falling configuration

5.6.2. Design and Implementation of Reversible Logic Multiplexer With Garbage Input/Output

The proposed reversible 2:1 multiplexer gate is realized using transistor implementation as described earlier. The required output can be obtained using only two transistors. However, three transistors are required for calculating garbage outputs. To obtain output G1, a pass transistor is used for passing the selection input S (select) to the output G1, when the selected input is 0, the PMOS transistor will be ON while NMOS will be OFF. Thus input A is passed to the output Y. If selection input is '1', then the NMOS transistor will be conducting while PMOS is in OFF state, passing input B to the output Y. Parallel to output Y, with input B, the output G2 is passed when S=0 (SEL) and input A, when S=1. This paper proposed reversible multiplexer gate in order to function as the 2:1 reversible multiplexer producing two garbage bits. The inputs are A, B and S. It depends on the preferred input S that the consequent information bit is accepted on to the output Y as shown in figure 5.11.

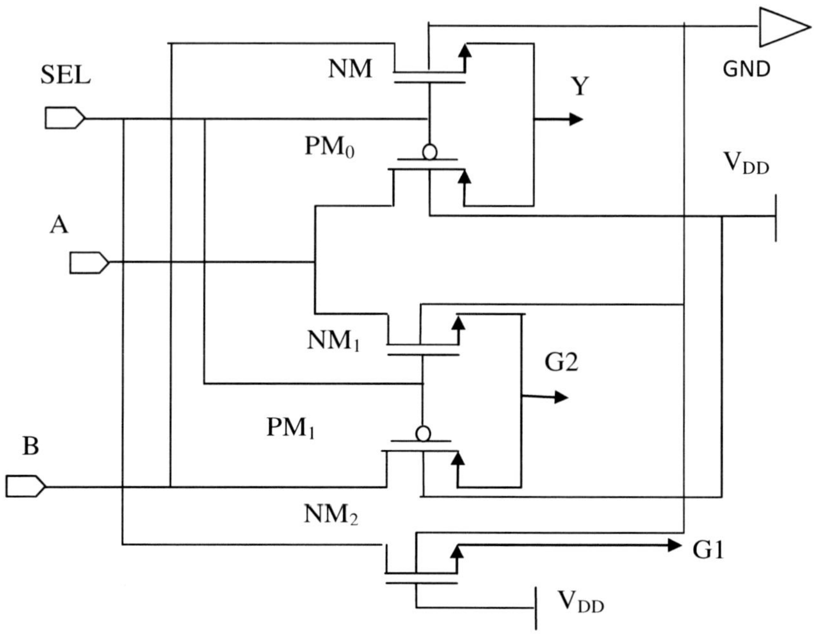

Figure 5.10. Transistor Implementation of 2:1 Reversible Multiplexer with Garbage Input/Output using GDI Cell Technique.

Table 5.2. Truth Table of 2:1 Reversible Multiplexer with Garbage Input/Output using GDI Technique

Input			Output		
SEL	A	B	G1	Y	G2
0	0	0	0	0	0
0	0	1	0	0	1
0	1	0	0	1	0
0	1	1	0	1	1
1	0	0	1	0	0
1	0	1	1	1	0
1	1	0	1	0	1
1	1	1	1	1	1

The proposed reversible logic gate circuit applying GDI cell technique reduces the area of the circuit, with less leakage power, delay and Power-Delay Product (PDP). This design of reversible logic circuit is requires the 2:1

multiplexer circuit where the number of garbage output is equal to two. If we design other reversible logic circuit like as 4:1.8:1 reversible multiplexer, it also generates the number of output i.e. six for 4:1 and 14 for the design of 8:1 multiplexer. The most cardinal thing for the design of reversible logic circuit is less number of transistor counts. Table 5.2 shows the transient response truth table for the GDI cell based reversible logic multiplexer and its implementation.

5.7. PERFORMANCE PARAMETER

Power Dissipation

In digital CMOS circuits, Power dissipation consists of four components as equated in equation (5.7).

$$P = P_{dynamic-switching} + P_{short-circuit} + P_{static-biasing} + P_{leakage} \tag{5.7}$$

Where P is the total power dissipation, $P_{dynamic-switching}$ is the dynamic switching power, $P_{short-circuit}$ is the short-circuit power, $P_{static-biasing}$ is the static biasing power and $P_{leakage}$ is the leakage power.

Dynamic switching power dissipation is caused by charging capacitances in the circuit.

The energy consumption during low-to-high output transition can be derived by integrating the

$$E_{vdd} = \int_0^{\infty} i_{vdd}(t) V_{dd} dt = V_{dd} \int_0^{\infty} C_L \frac{dv_{out}}{dt} dt = C_L V_{dd} \int_0^{V_{dd}} dv_{out} = C_L V_{dd}^2 \tag{5.8}$$

instantaneous power over the period of interest. Equation shows that it draws $C_L V_{dd}^2$ Joules from the power supply.

The charge stored on the load capacitor is equals to $C_L V_{dd}^2/2$ by equation. This means that only half of the energy supplied by the power source is stored in *CL*. The other half had been dissipated by the PMOS transistor. The high-to-low output transition dissipates the energy stored on the load capacitance into the NMOS transistor.

$$E_{C_L} = \int_0^{\infty} i_{Vdd}(t) V_{out}\, dt = \int_0^{\infty} C_L \frac{dv_{out}}{dt} v_{out}\, dt = C_L \int_0^{V_{dd}} v_{out}\, dv_{out} = \frac{C_L V_{dd}^2}{2} \quad (5.9)$$

To compute the power consumption, it is necessary to take into account how often the circuit is switched. Given a gate switching frequency *f*, the power drawn from the supply is given by:

$$P_{dynamic-switching} = C_L V_{dd}^2 f \quad (5.10)$$

Delay

The time constant τ is therefore the approximate time it takes for the output voltage to change after the input has changed. As such, the value τ is *approximately* the *propagation delay* tp of the CMOS inverter.

$$t_p \approx r_{DSP}\, C = \frac{C}{2K(V_{DD}-V_t)} \quad (5.11)$$

Sub-threshold Leakage Current

Sub threshold current occurs between drain and source when transistor is operating in weak inversion region, i.e., the gate voltage is lower than the V_{th}. The sub threshold leakage current for a MOSFET device can be expressed as:

$$I_{subthreshold} = I_0 e^{\frac{(Vgs-Vth)}{nV_T}} [1 - e^{-(\frac{V_{ds}}{V_T})}]$$

Where

$$I_0 = \frac{W\, \mu_0\, C_{OX}\, V_T^2\, e^{1.8}}{L}, V_T = \frac{KT}{q}$$

is the thermal voltage

V_{th}= threshold voltage

V_{ds} and V_{gs} = drain-to-source and gate-to-source voltages respectively.

W and L = effective transistor width and length, respectively.

C_{ox} = gate oxide capacitance

μ_0= carrier mobility and

n = sub-threshold swing coefficient.

Gate Oxide Tunneling Current

To control the short channel effects, oxide thickness must also become thinner in each technology generation. Aggressive scaling of the oxide thickness, in turn, gives rise to high electric field, resulting in a high direct-tunneling current through transistor gate insulator. The tunneling of electrons (or holes) from the bulk and source/drain overlap region through the gate oxide potential barrier into the gate (or vice-versa) is referred as gate oxide tunneling current.

Gate oxide tunneling current can be expressed as:

$$\mathrm{Igate} = W.L.A\left(\frac{Vox}{Tox}\right)^{2} e^{\left(\frac{-B\left(1-\left(1-\frac{Vox}{Tox}\right)^{\frac{3}{2}}\right)}{\frac{Vox}{Tox}}\right)}$$

Where W and L are the effective transistor width and length, respectively,

Where

$$A = \frac{q^3}{16\pi^2 h \emptyset ox}$$

$$B = 4\pi\sqrt{2m_{ox}}\,\frac{\emptyset ox^{\frac{3}{2}}}{3hq}$$

m_{ox} is the effective mass of the tunneling particle, *fox* is the tunneling barrier height, t_{ox} is the oxide thickness, *h* is *1/2*p times Planck's constant and *q* is the electron charge.

Band-to-Band Tunneling Current

The MOS transistor has two pn junctions – drain and source to well junctions. These junctions are typically reverse biased, causing a pn junction leakage current. This current is a function of junction area and doping concentration. When 'n' and 'p' regions are heavily doped, band-to-band tunneling (BTBT) leakage dominates the reverse biased pn junction leakage mechanism. A high electric field across a reverse biased pn junction causes a current flow through the junction due to tunneling of electrons from the valence band of the p-region to the conduction band of the n-region. Tunneling current occurs when the total voltage drop across the junction, applied reverse bias (V_{app}) plus built-in voltage (Ψ_{bi}), is larger than the band-gap. The tunneling current density through a silicon pn junction is given by:

$$J_{BTBT} = A\frac{EV_{app}}{E_g^{\frac{1}{2}}}e^{\left(-B\frac{E_g^{\frac{3}{2}}}{E}\right)}$$

$$A = \frac{\sqrt{2m^*q^3}}{4\pi^3h^3},\ \text{and}\ B = \frac{4\sqrt{2m^*}}{3hq}$$

m* is the effective mass of electron; Eg is the energy-band gap; Vapp is the applied reverse bias; E is the electric field at the junction; q is the electron charge; and h is 1/2p times the Planck's constant.

References

[1] A. Wang and A. Chandrakasan, "A 180mV FFT processor using subthreshold circuit techniques", *IEEE ISSCC*, pp.292-529, 2004.

[2] E. R. Hsieh and S. S. Chung, "The proximity of strain induced effect to improve the electron mobility in a silicon–carbon source–drain structrue

of n-channel metal–oxide semiconductor field-effect transistors," *Appl. Phys. Lett.*, vol. 96, no. 9, p. 093501, Mar. 2010.

[3] K. Akarvardar, C. D. Young, D. Veksler, K.-W. Ang, I. Ok, M. Rodger, "Performance and variability in multi-VT FinFETs using fin doping," in Proc. of International Symposium on VLSI Technology, Systems, and Applications (VLSI-TSA), pp. 1-2, 2012.

[4] K. Kim, K. K. Das, R. V. Joshi, and C.-T. Chuang, "Leakage power analysis of 25-nm double-gate CMOS devices and circuits," IEEE Trans. Electron Devices, vol. 52, no. 5, pp. 980–986, May 2005

[5] P. Zuber, M. Miranda, M. Bardon, S. Cosemans, P. Roussel, P. Dobrovolny, T. Chiarella, N. Horiguchi, A. Mercha, T. Y. Hoffmann, D. Verkest, and S. Biesemans, "Variability and technology aware SRAM Product yield maximization," in *VLSI Symp. Tech. Dig.*, Jun. 2011, pp. 222–223.

[6] E. Baravelli, L. Marchi, and N. Speciale, "VDD scalability of FinFET SRAMs: Robustness of different design options against LER-induced variations," *Solid State Electron.*, vol. 54, no. 9, pp. 909–918, Sep. 2010.

[7] K. Maitra, A. Khakifirooz, P. Kulkarni, V. S. Basker, J. Faltermeier, H. Jagannathan, H. Adhikari, C.-C. Yeh, N. R. Klymko, K. Saenger, T. Standaert, R. J. Miller, B. Doris, V. K. Paruchuri, D. McHerron, J. O'Neil, E. Leobundung, and H. Bu, "Aggressively scaled strainedsilicon-on-insulator undoped-body high- *k*/metal-gate nFinFETs for high performance logic applications," *IEEE Electron Device Lett.*, vol. 32, no. 6, pp. 713–715, Jun. 2011.

[8] J. B. Chang, M. Guillorn, P. M. Solomon, C.-H. Lin, S. U. Engelmann, A. Pyzyna, J. A. Ott, and W. E. Haensch, "Scaling of SOI FinFETs down to fin width of 4 nm for the 10 nm technology node," in VLSI Symp. Tech. Dig., 2011, pp. 12–13.

[9] Sentaurus Process User Guide, Version D-2011.09, Synopsys Co., Mountain View, CA, 2011.

[10] V. Moroz and M. Choi, "Strain scaling and modeling for FETs," *ECS Trans.*, vol. 33, no. 6, pp. 21–32, 2010.

[11] G. Eneman, N. Collaert, A. Veloso, A. De Keersgieter, K. De Meyer, and T. Y. Hoffmann, "On the efficiency of stress techniques in gate-last n-type bulk FinFETs," in *Proc. ESSDERC Tech. Dig.*, 2011, pp. 115–118.

[12] L. Donetti, F. Gámiz, N. Rodriguez, F. Jiménez-Molinos, and J. B. Roldán,"Hole transport in DGSOI devices: Orientation and silicon

thickness effects,"*Solid State Electron.*, vol. 54, no. 2, pp. 191–195, Feb. 2010.

[13] V. S. Basker, T. Standaert, H. Kawasaki, C.-C. Yeh, K. Maitra, T. Yamashita, J. Faltermeier, H. Adhikari, H. Jagannathan, J. Wang, H. Sunamura, S. Kanakasabapathy, S. Schmitz, J. Cummings, A. Inada, C.-H. Lin, P. Kulkarni, Y. Zhua, J. Kuss, T. Yamamoto, A. Kumara, J. Wahl, A. Yagishita, L. F. Edge, R. H. Kim, E. Mclellan, S. J. Holmes, R. C. Johnson, T. Levin, J. Demarest, M. Hane, M. Takayanagi, M. Colburn, V. K. Paruchuri, R. J. Miller, H. Bu, B. Doris, D. McHerron, E. Leobandung, and J. O'Neill, "A 0.063 μm2 FinFET SRAM cell demonstration with conventional lithography using a novel integration scheme with aggressively scaled fin and gate pitch," in *VLSI Symp. Tech. Dig.*, 2010, pp. 19–20.

[14] N. Xu, X. Sun, W. Xiong, C. R. Cleavelin, and T.-J. K. Liu, "MuGFET carrier mobility and velocity: Impacts of fin aspect ratio, orientation and stress," in *IEDM Tech. Dig.*, 2010, pp. 8.5.1–8.5.4.

[15] N. Xu, F. Andrieu, J. Jeon, X. Sun, O. Weber, T. Poiroux, B.-Y. Nguyen,O. Faynot, and T.-J. King Liu, "Stress-induced performance enhancement in ultra-thin-body fully depleted SOI MOSFETs: Impacts of scaling," in *VLSI Symp. Tech. Dig.,* 2011, pp. 162–163.

[16] N. Xu, B. Ho, F. Andrieu, L. Smith, B.-Y. Nguyen, O. Weber, T. Poiroux, O. Faynot, and T.-J. K. Liu, "Carrier mobility enhancement via strain engineering in future thin-body MOSFETs," *IEEE Electron Device Lett.*, vol. 33, no. 3, pp. 318–320, Mar. 2012.

[17] T. Toffoli., "Reversible Computing," *Automata, Languages and Programming,* Journal W. Bakker and J. van Leeuwen, ED. Springer Verlag, vol. 85, pp. 632-644, 1980.

[18] H. Bennett, "Logical reversibility of computation," *IBM Journal Research and Development,* vol. 17, no.6, pp. 525-532, 1973.

[19] Landauer, R., 1961. "Irreversibility and heat generation in the computing process*," IBM Journal Research and Development,* vol. 5 no.3, pp. 183-191, July 1961.

[20] M. P. Frank. "Introduction to reversible computing: motivation, progress, and challenges," *Proceedings of the 2nd Conference on Computing Frontiers, Italy and New York,* USA, pp. 385–390, May 2005.

[21] Vlatko Vedral, Adriano Bareno and Artur Ekert, *"Quantum Networks for Elementary Arithmetic Operations,"* vol. 1, pp. 102-109, Nov 1995.

[22] Perkowski. M, Jozwiak. L, Kerntopf. P, Mishchenko. A, Al-Rabadi.A, Coppola.A, Buller.A, Song.X, Khann.M.M.H.A., Yanushkevich.S, Shmerko.V, and Chrzanowskas-Jeske.M.2001 "A general decomposition for reversible logic," *Proceeding of 5th International Reed-Muller Workshop,* Starwille, pp. 119-138, 2001.

[23] Md. Saiful Islam, "BSSSN: Bit String Swapping Sorting Network for Reversible Logic Synthesis", *Journal of Computer Science,* vol. 1, no. 1, pp. 94-99, 2007.

[24] Mr. Devendra Goyal, Ms. Vidhi Sharma, "VHDL Implementation of Reversible Logic Gates," *International Journal of Advanced Technology & Engineering Research (IJATER),* vol. 2, no.3, pp. 157, May 2012.

[25] Jan M Rabaey, Anantha Chandrakasan, Borivoje Nikolic, *"Digital Integrated Circuits: A Design perspective,"* 2nd Edition, Prentice-Hall, Inc.

[26] Neil H.E. Weste, David Harris and Ayan Banerjee, *CMOS VLSI Design, A circuits and system perspective,3rd* Edition, Pearson Education, 2005.

[27] Y. Omura, S. Cristovoveanu, F. Gamiz, B-Y. Nguyen, - *Silicon-onInsulator Technology and Devices 14, Issue 4 - Advanced FinFET Devices for sub-32nm Technology Nodes: Characteristics and Integration Challangesll* 2009, pp.45-54.

[28] Ravish Aradhya H V, Praveen Kumar B V, Muralidhara K N "Design of Control unit for Low Power ALU Using Reversible Logic," *International Journal of Scientific & Engineering Research,* vol. 2, no. 9, pp.1-7, September-2011.

[29] Md. Saiful Islam, M. M. Rahman, Zerina Begum and Mohd. Zulfiquar Hafiz, "Low cost quantum realization of reversible multiplier circuit," *Information Technology Journal,* vol. 8, no. 2, pp. 208-213, Feb 2009.

[30] Rashmi S.B, Praveen B, and Tilak B.G., "Design of Optimized Reversible BCD Adder/Subtractor," *International Journal of Engineering and Technology (IJET),* vol. 3, no. 3, pp. 230-234, May 2011.

[31] Prathibha Rani T.R, H.R. Bhagyalakshm, M.K. Venkatesha, "Optimized Reversible carry select BCD adders using Reversible logic gates," *International Journal Computer Technology & Application (IJCTA),* vol.3, no.4, pp.1550-1556, July-August 2012.

[32] Mohammad Samadi Gharajeh and Majid Haghparast "On Design of a Fault Tolerant Reversible 4-Bit Binary Counter with Parallel Load," *Australian Journal of Basic and Applied Sciences,* vol.6, no.7, pp. 430-446, 2012.

[33] Praveen.B, Vinay Kumar S.B "2^n:1 Reversible Multiplexer and its Transistor Implementation," *International Journal of Engineering Science and Technology (IJCER),* vol. 2, no.1, pp.182-189, Jan-Feb 2012.

[34] Vivek V. Shende, Aditya K. Prasad, Igor L. Markov, and John P. Hayes, Fellow, IEEE "Synthesis of Reversible Logic Circuits," *IEEE Transaction on computer-Aided Design of Integrated circuits and system,* vol.22, no.6, pp. 710-722, June 2003.

[35] John P. Uyemura, "Introduction to VLSI Circuits and Systems," Wiley, John & Sons Inc, pp.656, 2002.

[36] Mohammadi, M. And Eshghi, M. "On figures of merit in reversible and quantum logic designs," *Quantum Information Process, Springer Science,* vol. 8, no. 4, pp. 297–318. 2009.

[37] Praveen B, Rashmi S.B, Tilak B.G, "Transistor Implementation of Reversible PRT gates," *International Journal of Engineering Science and Technology (IJCER),* vol. 3, no. 3, pp. 2289-2297, 2011.

[38] Kaushik Roy, fellow, IEEE, Saibal Mukhopadhyay, student member, IEEE, and Hamid Mahmoodi –Meimand student, IEEE "Leakage Current Mechanisms and Leakage Reduction Techniques in Deep Submicrometer CMOS Circuits," *Proceedings of The IEEE,* vol. 91, no. 2, pp.305-327, February 2003.

[39] S. A. Pallav Gupta, Student Member, IEEE, Abhinav Agrawal, and Niraj K. Jha Fellow, IEEE "An Algorithm for Synthesis of Reversible Logic Circuits," *IEEE Transactions on Computer-Aided Design of Integrated Circuits and Systems,* vol. 25, no. 11, pp. 2317-2330, November 2006.

[40] Praveen.B, Vinay Kumar S.B "2^n:1 Reversible Multiplexer and its Transistor Implementation," *International Journal of Engineering Science and Technology (IJCER),* vol. 2, no.1, pp.182-189, Jan-Feb 2012.

AUTHORS CONTACT INFORMATION

Dr. Shyam Akashe
Associate Professor
Electronics & Communication Engineering
ITM University
N.H-75, Jhansi Road, Turari, Gwalior-474001 (M.P.)
Mobile No.: 094251-46040
E-mail: Shyam.akashe@itmuniversity.ac.in

Dr. Khusbou Mishra
Assistant Professor
Electronics & Communication Engg. Department
GEC, Gwalior
Telephone: +91-9926664564
Email: khushboomishra.vlsi@gmail.com

INDEX

#

20th century, 3

A

abstraction, 39
adaptation, ix, 96
ADC, 59
advancement, 42
algorithm, 13, 23
alters, 5
arithmetic, 12
atoms, 6
automation, 62, 63

B

band gap, 10, 119
bandwidth, ix, 17, 102, 104
Bangladesh, 84
base, 3, 4, 54, 110
behaviors, 12
bending, 49
bias, ix, xiii, 10, 52, 54, 57, 85, 87, 90, 92, 94, 95, 96, 98, 106, 119
building blocks, 12

C

candidates, 102
carbon, 6, 83, 119
carbon nanotubes, 6
categorization, 102
challenges, 8, 23, 34, 83, 94, 121
channel length, vii, 1, 9, 47
circuit delay, vii, 1, 9, 21
clustering, 97
CMOS technologies, vii, 1, 10
commercial, 3
communication, viii, 17, 65
communication systems, 17
compensation, 95
competition, 4
Complementary Metal Oxide Semiconductor, viii, xvii, 4, 38, 65
complexity, 5, 21, 113
computation, 23, 24, 25, 45, 75, 83, 86, 103, 121
computational performance, 4
computer, viii, 3, 6, 7, 23, 24, 65, 76, 102, 104, 123
computing, vii, 1, 12, 23, 25, 27, 33, 34, 37, 39, 75, 83, 84, 103, 104, 121
conduction, 2, 19, 30, 103, 119
conductivity, 3
conductor, 47
conference, 62, 63

configuration, 22, 45, 46, 111, 114
Congress, 33
construction, 23
consumption, vii, viii, ix, xii, 3, 4, 5, 8, 13, 15, 20, 37, 38, 39, 40, 42, 43, 44, 45, 49, 51, 54, 57, 65, 69, 70, 85, 89, 94, 101, 104, 107, 117
convention, 38
cooling, 5, 40
copper, 2, 17
correlation, vii, 1, 9
cost, 3, 5, 7, 23, 40, 55, 60, 66, 76, 84, 112, 122
CPU, xi, 7
cryptography, 23

D

data communication, 16
decomposition, 13, 15, 31, 32, 84, 122
decoupling, 22
defects, 57
deficiency, 47
degradation, viii, 20, 85, 86, 102
designers, vii, 37, 38, 39
destruction, 52
detection, 34
DFT, 34
dielectric constant, 9
diffusion, 22, 46, 47, 48, 110
digital communication, 25
distribution, xiii, 9, 88
doping, 48, 57, 83, 95, 102, 119, 120

E

editors, 60
electric field, 44, 48, 49, 118, 119
electrical conductivity, 2
electron(s), 2, 3, 48, 49, 83, 102, 118, 119
electronic systems, 4
energy, 23, 24, 25, 75, 94, 98, 99, 100, 103, 116, 117, 119
energy consumption, 94, 116
energy efficiency, 99
engineering, 3, 10, 23, 121
environment, 40, 58
epitaxial growth, 103
exertion, 38

F

fabrication, 3, 7, 8, 16, 57
families, 46, 111
fault detection, 34
FFT, 83, 99, 119
fluctuations, 57, 95
force, 5, 51, 102
formula, 21, 42, 55
France, 33

G

garbage, viii, ix, xii, xiii, 27, 28, 37, 39, 46, 65, 77, 79, 80, 81, 82, 101, 104, 111, 113, 114, 116
Gate -Diffusion -Input, ix, 101
GDI cell, ix, xv, 45, 101, 110, 111, 115
germanium, 3, 4
Germany, 33
global competition, 5
grounding, 47
growth, 5

H

heat removal, 5
height, 45, 103, 119
history, 2, 30
human, 47
human body, 47
hybrid, 106

I

identity, 3
image, 8

immunity, 94
India, 33
industry(s), 2, 3, 4, 5, 6, 7, 9, 10, 40, 44
input signal, 3, 17, 22
insertion, 50
insulators, xv, 9
integrated circuits, viii, 7, 8, 16, 37, 39
integration, 16, 121
integrity, 57, 102
interface, viii, 85
inventions, 3
inversion, 21, 49, 103, 117
investors, 6
irreversible logic Multiplexer, vii
Islam, 34, 60, 84, 122
issues, viii, 4, 34, 40, 42, 44, 58, 76, 85, 89, 98, 104, 110, 112
Italy, 34, 83, 121

L

lead, 44, 92, 102
leakage current, vii, viii, ix, 1, 8, 9, 22, 25, 44, 45, 47, 48, 50, 51, 52, 65, 85, 86, 88, 89, 90, 91, 102, 104, 106, 108, 117, 119
leaks, 52
learning, 23
lithography, 121
logic gate diagrams, vii, 1
logic synthesis, vii, 1, 22, 23, 34, 75, 103
logical operation, vii, 1, 24, 25, 26, 75, 103

M

magnetic field, 2
magnitude, 86, 94
major issues, ix, 101
majority, 23
management, 98
manipulation, 23
manufacturing, xv, 4, 5, 8, 9, 16
mapping, vii, 1, 23, 24, 25, 75, 76, 103, 112
mass, 119
materials, 2, 3
measurement(s), 47, 95, 98, 100
memory, viii, xi, 7, 11, 12, 52, 65, 94
methodology, 8, 20
microelectronics, 6
micrometer, 61
migration, 8
mission, 30
mobile applications, vii, 37, 38
mobile phone, 5, 98
models, 56, 58
motivation, 34, 83, 121
multiplier, 13, 33, 34, 60, 79, 84, 122
MUX circuit, vii, 1

N

nanometer, 50
nanotechnology, viii, 5, 24, 25, 28, 34, 46, 65, 76, 77, 104, 111
next generation, 7
nodes, 8, 21, 22, 40, 102

O

obstacles, 8, 110
off-current, vii, 8, 10
one-to-one mapping, vii, 1, 23, 24, 25, 75, 76, 103, 112
operating frequency, vii, 1, 9
operations, 24, 26, 75, 103, 104, 106
optimization, viii, 13, 23, 47, 76, 85, 98, 112
output values, vii, 1, 23, 75, 82, 103
overlap, 118
ox, 9
oxide thickness, 8, 9, 44, 48, 49, 95, 118, 119

P

parallel, ix, 8, 11, 55, 70, 84, 101, 106, 107, 108
PDP, ix, 56, 79, 82, 102, 104, 111, 115
PGD, 56

physics, 3, 30
pitch, 5, 121
PM3, 92
power consumption, vii, viii, ix, 3, 4, 5, 13, 15, 20, 37, 38, 39, 40, 42, 43, 44, 45, 49, 51, 57, 65, 69, 70, 85, 89, 94, 101, 104, 107, 117
power dissipation, viii, ix, xii, 5, 21, 38, 39, 40, 41, 42, 43, 44, 46, 50, 59, 68, 85, 101, 111, 116
power lines, 16, 50, 70, 72, 105, 109
power reduction, vii, 1, 23, 49, 52, 54, 55, 75, 88, 89, 97, 103
Power-Delay Product, ix, 102, 115
predictability, 10
principles, 59
probability, 15, 48
propagation, 31, 32, 55, 71, 104, 107, 117
protection, 47
purity, 3

Q

quantum computer, 23, 24, 76, 84, 104
quantum computing, 24, 25, 76, 104
quantum networks, 24, 76, 104

R

reality, 25
recovery, 24, 75, 104
rectification, 2
regulations, 94
relative size, 49
reliability, 4, 10
requirements, vii, viii, 21, 37, 39, 65
researchers, vii, 37, 38
reserves, 39
resistance, 2, 50, 55, 56, 57, 63, 73, 103
resolution, 8
resource allocation, 13
response, viii, 10, 25, 37, 39, 79, 82, 111, 116
reversible circuit, vii, 1, 9, 10, 23, 28, 34, 44, 76, 112
reversible gates, vii, 1, 23, 27, 28, 46, 75, 76, 77, 84, 103, 111, 112, 113, 114
reversible logic, vii, viii, ix, xi, 1, 5, 8, 9, 10, 20, 23, 24, 25, 27, 28, 33, 34, 35, 37, 39, 40, 41, 43, 46, 51, 55, 75, 76, 77, 79, 82, 84, 86, 101, 103, 104, 111, 112, 113, 115, 122
reversible logic circuit, vii, ix, 1, 5, 9, 23, 24, 25, 40, 41, 75, 76, 77, 82, 101, 103, 104, 112, 114, 115
reversible logic Multiplexer, vii, 1
risk, 52
room temperature, 24, 75, 103
root, 26, 77, 113

S

saturation, 56, 71, 107
savings, 54
scaling, vii, ix, xi, xv, 4, 5, 6, 7, 8, 9, 10, 20, 22, 23, 37, 39, 40, 44, 47, 50, 54, 55, 57, 63, 73, 99, 102, 104, 109, 110, 118, 120, 121
Schrödinger equation, 103
scope, 3
self-control, 90
semiconductor(s), 2, 3, 4, 5, 7, 8, 9, 10, 19, 30, 40, 44, 45, 63, 83, 120
sensing, 40
sensor network, 68, 102
servers, 30
shape, 48
showing, 12
signals, viii, 11, 15, 17, 19, 24, 46, 66, 72, 75, 85, 103, 104, 106, 109, 111
silicon, 3, 4, 46, 58, 63, 83, 102, 104, 119, 120
silver, 2
simulation(s), 8, 10, 22, 46, 58, 79, 82, 94, 102, 111
Singapore, 30
SiO_2, 9
software, 102

SOI, 46, 58, 111, 120, 121
solid state, 4, 12, 23, 60, 83
solution, 73, 109
Spain, 33
state(s), 11, 12, 17, 23, 24, 42, 55, 67, 76, 80, 89, 98, 104, 114
statistics, 89
stress, ix, 10, 57, 101, 102, 120, 121
stressors, 102
structure, xii, 4, 13, 57, 70, 86, 94, 102, 107
style(s), v, vi, vii, 1, 13, 17, 21, 22, 68
substrate, xii, xiii, 3, 52, 53, 54, 87, 90, 92, 96, 103
Sun, 121
suppression, 57, 60
susceptibility, 57
synchronization, 23
synthesis, vii, 1, 22, 23, 34, 63, 75, 76, 103, 112

T

target, 31, 47, 58
techniques, vii, viii, 1, 7, 10, 16, 37, 39, 42, 46, 47, 60, 61, 83, 85, 86, 89, 90, 97, 99, 111, 119, 120
technology(s), vii, viii, xi, 1, 3, 4, 5, 6, 7, 8, 9, 10, 13, 16, 19, 20, 23, 24, 25, 30, 37, 38, 39, 40, 44, 47, 50, 52, 54, 57, 61, 65, 69, 73, 75, 82, 83, 85, 86, 95, 96, 97, 98, 102, 103, 104, 109, 118, 120
temperature, 2, 10, 24, 40, 75, 95, 103
temporal variation, 95
terminals, 3, 41, 72, 106, 108
testing, 13
thoughts, vii, 37, 38
titanium, 57, 63
top-down, 45
topology, ix, 65, 101
trade, 5, 13
trade-off, 5
transactions, 61
transformation, 23
transmission, viii, ix, 13, 19, 20, 45, 65, 69, 70, 78, 101, 107
transport, 120
tunneling, 8, 9, 48, 49, 118, 119

U

United States (USA), 31, 32, 33, 34, 83, 121
UV, 8

V

vacuum, 3, 4
valence, 119
variables, 27
variations, ix, 47, 57, 85, 95, 110, 120
vector, 24, 26, 76, 112
velocity, 56, 121
Very Large Scale Intigration, viii, 65
VLSI, viii, xvii, 3, 4, 5, 14, 18, 19, 20, 25, 32, 33, 34, 38, 40, 44, 46, 60, 61, 62, 63, 64, 65, 83, 84, 89, 98, 99, 100, 111, 120, 121, 122, 123

W

Washington, 33
wells, 52
wires, 4
Wisconsin, 59

Y

yield, 31, 83, 95, 110, 120